Praise f

Dr Catherine Ball doesn't ju
potential solutions, she brings uiciii ···
that's impossible to ignore.
Professor Brian Schmidt AC, Nobel Laureate; Vice-Chancellor and
President, Australian National University

From Big Data to FrankenFoods, from Spacecations to the Metaverse,
Catherine Ball takes us on a wild ride into our future.
Adam Spencer, broadcaster and University of Sydney Ambassador

Finally! A book that shows us the opportunities that tech and humanity can
embrace, rather than the battle of who wins. Catherine takes us on a broad and
deep tour of the rich tapestry that might be the future – if we dare to believe.
Dom Price, work futurist, Atlassian

Catherine has a unique ability to synthesise the advances in technology into
a compelling vision of what our future could hold.
Kylie Ahern, CEO and Founder, STEM Matters

Who better to predict our future than someone inventing it?
Professor Toby Walsh, UNSW Laureate Fellow, world leading researcher
in artificial intelligence and author of *Machines Behaving Badly*

For those of us not interested in toys for boys, this is a compelling book about
the brilliant technology we all really need for a better future.
Robyn Williams AO, host of ABC Radio National's *The Science Show*

A provocative and intellectual force, Dr Cath hands us confronting facts about the
future and asks the greatest question of all – what part do YOU want to play?
Andrea Clarke, author of *Future Fit*

In an increasingly complex and often disturbing world, the sensible calls to action
in this book remind us there are achievable ways of taking greater responsibility for
steering towards a more positive collective future.
Professor Katherine Daniell, School of Cybernetics, Australian National
University; Chevalier, Ordre national du Mérite

I couldn't put this book down, and I'm sure you will feel the same once you
open to the first page. Thanks, Cath, for sharing your brain!
Professor Sarah Pearson, Non Executive Board Member, Global Innovation Fund;
Member of ANU Council

Converge is a shocking but necessary reminder of how asleep the world is, coupled with a motivational and clear way to wake it up. I'll be thinking about *Converge* for a long time to come.
Sarah Bartholomeusz, author and entrepreneur, You Legal Pty Ltd

A fantastic and timely book by one of the industry's premier leaders.
Adjunct Professor Alex Antic, RMIT; Executive Board Member, Global AI Ethics Institute

Dr Catherine makes complex topics easy to understand. Water-cooler conversation is about to get 100 times nerdier.
Amanda Johnstone, CEO of Transhuman

This is a rollicking read from start to finish. Catherine relates an optimistic future despite the challenges of complexity, uncertainty and disruption.
Associate Professor Sarah Kelly OAM, Marketing & Law, University of Queensland Business School

Converge is a much-needed reminder that while technology will change the future, exactly how it does that is up to each of us.
Lorraine Finlay, Human Rights Commissioner, Australian Human Rights Commission

This is powerful, purposeful and written by a proven futurist standing tall to guide us safely to what is conceivable in our ever-evolving future.
Yasmin Grigaliunas, Founder, World's Biggest Garage Sale and Circonomy

Dr Catherine Ball is a true storyteller, and it's refreshing to read her perspectives on how we may just need to look at things a little differently to make immediate and permanent positive change.
Vanessa Garrard, serial entrepreneur, EY Australian Entrepreneur of the Year

Converge carves out the human-centred future where technology augments and enables the human experience. The future is not scary; it's diverse and exciting.
Anne-Marie Elias, Fellow, Innovation & Entrepreneurship, University of Technology Sydney

The future is already here. If you want to know what it looks like, *Converge* is an excellent place to start.
Mark Pesce, author and futurist

Catherine Ball

CONVERGE

A futurist's insights into
the potential of our world as
technology and humanity collide

Dr Catherine Ball

MAJOR
STREET

To my Auntie Arlene, who I miss every day.

Thank you for being the best a human can be – a protector of children, a guide, a warrior, an example to me. Thank you for always being there, and for helping me create the future I wanted for myself.

MAJOR
STREET

First published in 2022 by Major Street Publishing Pty Ltd
info@majorstreet.com.au | +61 421 707 983 | majorstreet.com.au

© Dr Catherine Ball 2022
The moral rights of the author have been asserted.

A catalogue record for this book is available
from the National Library of Australia.

Printed book ISBN: 978-1-922611-52-9
Ebook ISBN: 978-1-922611-53-6

Cover design by Typography Studios
Internal design by Production Works
Printed in Australia by Griffin Press, an Accredited ISO AS/NZS 14001:2004
Environmental Management System Printer

10 9 8 7 6 5 4 3 2 1

Disclaimer

The material in this publication is in the nature of general comment only, and neither purports nor intends to be advice. Readers should not act on the basis of any matter in this publication without considering (and if appropriate taking) professional advice with due regard to their own particular circumstances. The author and publisher expressly disclaim all and any liability to any person, whether a purchaser of this publication or not, in respect of anything and the consequences of anything done or omitted to be done by any such person in reliance, whether whole or partial, upon the whole or any part of the contents of this publication.

Contents

'It is never too late to be what you might have been.'
– Attributed to Mary Ann Evans, aka George Eliot
(a fellow writer from Nuneaton)

Foreword

I grew up in Iran, watching *Star Trek* dubbed in Farsi. I loved the tagline 'to boldly go where no one has gone before'. I loved the show because it portrayed a society that had all its needs met and now was in pursuit of knowledge and exploration. I thought to myself, 'This is the future of humanity'.

What I saw on the TV screen was a vast contrast to what I saw outside my window. The world was not just or equitable; food and basic needs were not easily obtained; and if you got sick or broke your arm, it would take months to heal or get better, if at all. But I was a hopeful child with a curious and imaginative mind, and I would fill my free time with reading my Golden Series book on Madame Curie and invention of the X-ray machines, or Jules Verne's *From the Earth to the Moon* or *Twenty Thousand Leagues Under the Sea*. In my child's mind, the world of *Star Trek* was all possible and around the corner.

Science was core to my world – it was how I would find answers to thousands of questions that always swirled in my head: 'Why is the sky blue?' 'What are stars made of?' 'Are we alone in the universe?' 'How did the universe came to be?' 'Why are we here?' The questions never stopped. Some were answered, and some are still unanswered questions swirling in my mind.

As I grew up in the 1970s, pace of change seemed much slower that what I imagined it would be. The world around me was changing, but not in a significant way. The world was not going in the direction of the *Star Trek* world I had fallen in love with. Instead, we had war, death, disease and famine.

As I immigrated to the US and became an engineer, and a tech entrepreneur in the 1980s and 1990s, I was more convinced than ever that we can solve all sorts of problems and answer so many questions with the tools that science provides us. The world around me was changing but it was still far from that *Star Trek* utopia. As an entrepreneur, I was able to look into the future and get inspiration from it. I would imagine and build solutions and technologies to move us one step closer to that future.

It was during this time that I became involved with the XPRIZE foundation and, with my family, sponsored the first competition that catalysed the creation of a whole new commercial space economy and a new era of access to space for humanity. I got my chance to fly to the International Space Station in 2006 and see our world from a vantage point that only about 500 human beings have experienced. That vantage point brought all those questions back, swirling in my mind, but this time with a sense of belonging to a special place in our universe: our home planet, Earth. It also made it clear in my mind that if we want to explore out there, we need to fix our problems down here.

My serendipitous collision with XPRIZE gave me an opportunity to do just that – to help fix our problems down here, while opening our path to explore out there. The mission of XPRIZE was my mission: to build a hopeful and abundant future for *all*. It was to build my *Star Trek* future. As a tech entrepreneur, I knew the solutions were out there, in every corner, in the minds and hearts of people living their lives, not thinking about the problems or believing that they could solve them. XPRIZE was the platform to focus this cognitive energy of the world on solving the most important problems we face, with a hopeful vision toward a desired future of our own making.

Dr Catherine Ball is one of those visionary futurists in our XPRIZE network of experts, helping to envision this desired future. She understands full well the positive power of technology and the potential destructive uses of it, as she describes both sides of the coin in many different industries and technologies in this book. In today's fast-changing world, where convergence of many technological changes is happening at an exponential rate, the future is unpredictable and, for some, even incomprehensible. It is only our belief in the goodness deep in the heart of humanity that helps us look beyond our past mistakes, and the greed that has driven our world toward the cliff of existence, and know that we can choose differently.

Converge is an invitation to be curious and open-minded; to understand and explore all the changes that are happening around us in our world, and their potential; and to choose to build a better future. It is an invitation to care about and demand change in the way our systems look at and value risk. It is an invitation to put humanity and its long-term wellbeing at the center of our future designs, instead of short-term profit. It is an invitation to see the potential we have and to go where no one has gone before...

Anousheh Ansari
CEO, XPrize

The future is already here...

... and it's the end of the world as we know it.

I am a human 3D printer: I make people. (I have two sons.) I am a creator and consumer, a resource user – and a privileged one at that. With current technologies, the rate of my consumption is more than one planet Earth can handle. And there is no Planet B.

I am also a scientific futurist. From a drone's-eye view, I see endless potential every day. I believe the future flies. The future is interconnected. The future is powered by purpose. The future is an incredible place.

The future is convergence. It's characterised by fast combining, evolving and hybridising technologies. There are patterns and predictions being touted as prescience, but I most enjoy seeing the complex made simple – the human applications of technology that demonstrate the very reason technology exists.

To think like futurists, we must find our inner Yoda: 'Always in motion, the future is'. We must unlearn what we have learned about the world and allow our brains to dream of what the future might be, without the constraints of the current reality holding us back. We must empty our knowledge cups so that we can process new ways of thinking and novel information, free from the bias of our experiences.

How do we find the precarious balance between technological capability, ethical obligation and fiduciary responsibility? Our current global economic systems place value on things that our grandchildren will have to pay for and our planet can't sustain. We are the only species that knowingly spoils its own nest, and we seem unable to break free from our addiction to consumption.

Separated by 150 years, two blokes – Thomas Robert Malthus and Robert Solow – had very different thoughts about how our society would collapse (or not). Malthus was a demographer and economist whose staple concept of the self-destruction path of Homo sapiens was the 'Malthusian trap': humans had conquered agriculture and were able to grow enough food to feed themselves, but that led to population growth, which in turn created a higher demand for food. That then led back to famine from feast, eventually causing inequality and poverty. Eventually, he theorised, we humans will outgrow our capability to produce enough food, and society will collapse when Mother Nature can't give or take any more.

On the other hand, economist Solow's nonlinear economic model argues that technological advances will always keep us free from the Malthusian trap. In other words, humans will constantly advance technologically in order to outpace any potential threat to our survival. When Solow created his more optimistic model in the 1950s, he couldn't have possibly dreamed of the technologically advanced and driven society that we currently live in, but he guessed something like it was coming.

In the end, it comes down to what we want to believe. Is our species going to overpopulate and destroy our planet? Will Mother Nature fight back? Will we have an opportunity to make different decisions? What do we need to change so that every human can reach their full potential while using only a sustainable allocation of resources?

Human crises circulate, like fashion. The COVID-19 pandemic is flavour of the month as I write this. Everyone likes to refer to the last biggie, the 'Spanish flu' (which was actually first detected in Kansas, USA), and to the economic recovery in the roaring 1920s and all the hope and life-affirming growth that came with that. Now we are in the 2020s and in a place we never thought we would be in again: a microbe has brought our health systems, economies and societies to their knees. Were we arrogant to think that this wouldn't happen again? With our climate having changed, and with our war on biodiversity and wild spaces, we are actually more at risk from pandemics now than we have ever been. How can we plan for what is possible? Well, that is the kind of conversation this book has been written to cultivate at dinner parties, around the water cooler and on group video calls.

The future is already here – it's just in pieces. The privileged societies of the West have early-bird access to it: from medicine to food, travel and technology. Equity and equality are ever-growing issues as the digital divide widens. If we don't use all our human ingenuity, we are never going to get ourselves out of the dilemmas we face.

In the face of an increasingly commercial and technology-driven world, we need to start doing things that only humans can do. We are really good at creativity. We can develop ideas about the way we want to live our lives, and how we might use technology and other advances to become healthier and more fulfilled. We don't want to sit like robots in front of computers, typing away in mundanity without curiosity. Let the robots do the robotic stuff, and let us be more human.

Change in business culture is slow. 'This is the way we have always made profit'; 'This creates a low risk for shareholders'; 'This has been assessed by a risk committee' – these are the types of statements used to justify expensive and short-sighted economic management methods. Business staples won't change without us generating the need or desire for them to change. There is an unpredictable, intangible and unmodellable opportunity cost in not taking advantage of new and emerging technologies and the opportunities they bring. However, unless there is a tangible driver – such as workplace health and safety, or the risk of fines for inaction – there is little to spur change in how business operates. The climate emergency and the spate of legal actions that have commenced against big corporations regarding their carbon footprint has been an interesting start to the next wave of change as we move from the fourth industrial revolution to the fifth.

There will always be gnarly, seemingly impossible problems gnawing at the belly of all we hold dear. There is always an unjust war waging somewhere; there are myriad threats of ecosystem damage, acceleration of species extinction, destruction of wild places by ferocious fires, overconsumption of resources, the plastic problem, overfishing, melting ice and increased demands on finite resources.

But it isn't too late.

We can turn this around with the Solovian approach of nonlinear progress, utilising technology to save us from the brink. Nothing in nature fits neatly into a straight line. When looking to predict the future, we have to think about levers and tipping points rather than a linear continuum. We need to ask what's possible, what's probable and what's preferable.

We can write our own future, to a certain extent. What's preferable to many may be poisonous to others. That is the beauty and the curse of democracy – we allow for differences of opinion and for outliers. But the numerous problems we are facing are not insurmountable if we solve them together. Inequality should be a thing of the past, but it's hard to convince people that they need to 'decolonise' their thinking. Our systems aren't perfect – we can always do better. Innovation starts with conversation, and a reminder that we have two ears and one mouth. Wouldn't the world be a better place if we listened to understand rather than respond?

This is where you come in, dear reader. With fake news often overpowering science communication in the media, many people are craving insight into the overwhelming rise of technology around them. There is a global issue around literacy in science, technology, engineering and mathematics (STEM). This really needs to change – fast. Everyone deserves a chance to get educated, get involved and help curate technology, and to contribute to the conversation around how to adopt it and what we should be wary of. Picking up this book is a great start.

In the following chapters, you'll be taken on a roller-coaster ride into the future. Keep your arms inside the carriage at all times. Whether you view it as a terrifying ghost train or a gravity-defying thrill ride is really up to you. Technology is neither good nor bad – it is how it is applied that matters. How we live, the choices we make, how we want our politicians to represent us, and the manner in which we regulate the companies making use of the future is all up to us. We have choices about how we consume, how we communicate and how we determine what is valuable.

All my written works on the future are love letters to my sons, because the future is theirs. I just hope we leave them a good one.

A mind of one's own

The future of intelligence

'I am so clever that sometimes I don't understand
a single word of what I am saying.'
– Oscar Wilde, 'The Remarkable Rocket'

When technology is invisible to you, it is probably working. If your smartphone, your fridge, your electric toothbrush and the sensors in your car work, you barely notice them. But when technology breaks down, doesn't play nicely, is not fit for purpose, or doesn't predict or autocorrect the way you want it to, suddenly you see the cracks. It's then that you see technology for what it is: another tool, another machine that needs fixing.

The ability to fix something when it is broken is the true measure of our competency. Have you ever built or fixed your computer? Ever rewritten software on your smart device? The vast majority of us have become consumers of technology, but not creators of it. Because we are not creating technology, we can't understand how it is constructed. If we can't understand its construction, how can we ever fully comprehend its potential uses and limits? For many of us, it's been years since we were formally educated – and chances are the basics of emerging technologies were not part of our schooling. We are starting from ground zero, and this means we are reliant on the people who *do* build these systems to be honest with us about what they are and how we should relate to them.

The smart device you use to make calls, take photos, post on social media and check your bank balance is likely your most common interaction with robotics and artificial intelligence (AI). AI is simply intelligence demonstrated by machines that can perceive their environment and take actions that maximise their chances of achieving goals. Your smart device is an example of convergent technology – it takes multiple methods of data collection and delivery, and puts it all in one place. Convergence (and divergence) are key to future technologies.

Do you feel educated, informed and confident about how you use that smart device? What about the other technologies in your life? Here is the key: you have the ability to give consent to the technologies that you use. The machine can still be unplugged if you really want to switch it off. But do you *want* to switch it off, or have you found it meets a need in your life that you never knew you had before?

Are you in charge of your own mind?

'I think, therefore I am', philosophised René Descartes in the 17th century. What would he think today? How much of our thinking is really ours, and how much is being steered by the stimulation we are bombarded with on a daily basis?

Have you ever bought something recommended by an algorithm, or just let the next suggested video play? YouTube has been lauded for its accurate predictive algorithms that busily work away to find you the next clip you'll like. The aim, of course, is to keep you watching, to keep you on the platform, to sell more advertising – to make more money off you without taking any money off you, just by keeping your eyes on the screen for a few minutes longer.

As a new parent, I was part of many discussions about limiting screen time for kids. I learned about how bad screens are for developing humans – for their eyes, posture, language skills, behaviour and even the neurons forming in their brains. And yet we, as adults, are spending more and more time in front of screens, often barely questioning what it's doing to us.

Have you ever wondered what 'free will' is in a world filled with technology that probably knows more about you than you know about yourself? How can you still make choices when algorithms have learned all of your preferences and can predict with accuracy what you'd like to eat for dinner, how you'll vote, your sexual preference and your clothes size? How can we fight the coming storm of mundanity where anything vaguely spontaneous seems contrite and predictable? Are we satisfied with giving away information about ourselves for free to tech billionaires? Do we value the next face-changing app more than the risk to our cybersecurity?

The dopamine factor

Dopamine is the boss. It is a chemical produced by the brain when we feel a sense of pleasure or reward. It provides the feel-good hit when someone likes our post, or even when an app notifies us about something mundane and completely unrelated to us. Yep, that's right: that annoying app that sends you a notification for no particular reason is actually triggering brain chemistry that can be addictive. We click on links because we feel rewarded by it.

Games, apps, social media, dating sites, banks and news outlets are all interacting with our brains in ways and at frequencies never seen in the past. This is why people are now recommending 'slow reading'. Similar to how 'slow food' is good for our bodies, the idea is that reading from paper-based sources rather than screens is actually good for our brains – and our long and strong neurons, which create longer-term memory, recall and reason – and therefore potentially good for our mental health.

When I think I have misplaced my mobile phone, I get very anxious about where it could be. There's a name for this phenomenon: nomophobia. If I actually lost my phone, it could be easily replaced, because I have everything backed up to the cloud – yet I am still anxious. What drives this anxiety? It seems we can't deal with not touching our phones. Studies have shown that even having your phone in front of you at dinner can reduce your social interactivity and create the urge to check your phone more frequently than if you had it away in your bag. There's even a new word to describe what happens when you choose checking your phone over speaking with your partner – 'phubbing' – yet doing this has been shown to create emotional distance, leading to relationship breakdown. Other studies highlighted by the media (to make parents feel even worse about themselves) have questioned whether parents being addicted to their phone screens has slowed their children's development in areas such as smiling, interaction, self-regulation and emotional connection. Horrific, really, and yet even I was a mother who looked at my phone while breastfeeding.

The 'still face' experiment developed in the 1970s clearly demonstrated that if a child doesn't get that interactivity from their carer, they will try anything they can to reconnect. Eventually the child becomes distressed and unable to cope with the lack of facial interactivity. What does this mean for us and the technology we are spending all our time on? Are we accidentally denying our loved ones our attention as we get sucked into the scroll-vortex of social media? Are we choosing to damage our relationships and cause distress to our children by spending too much time looking at our screens, or are we addicted?

Nicholas Carr's book *The Shallows: What the internet is doing to our brains* includes real-world examples of how technology that was touted as helpful is actually harming humans. Carr says short information in means short information out. When we become addicted to hyperlinks and information that hops around like a rabbit, this trains our brain to forget what we've read almost immediately.

Carr's book helped me identify the effects of my own digital addiction. For example, I recognised how much more difficult it had become for me to concentrate on reading a book. I couldn't seem to find digital peace – I constantly checked the news and social media. Even writing this book has required me to try to eliminate all distractions. With two kids under five, it was certainly interesting, but it also made me remember how I worked when I was writing my PhD thesis – my most productive hours were 8 p.m. to 2 a.m., when fewer people were contacting me. Yet now I have people in the US and Europe interacting with me when everyone in Australia is asleep.

Sometimes I think my phone has vibrated in my pocket when it actually hasn't. Sometimes I think I hear a notification when there hasn't been one. I am constantly anxious about missing a call or text. As a new mother, I still sometimes have phantom crying anxiety, where I think I have heard my baby cry, but they didn't. Why does my brain think it hears my children? Evolution. Why does my brain think it hears my phone? The same reason.

One time, after spending hours doomscrolling through Instagram, I lay in bed and couldn't even remember what I had just looked at. When we aren't actively choosing what content to consume, it's like chewing gum for the brain: it leaves no mark and provides no 'nutrition'. It was a pointless, futile use of brain power when I could have been reading, working, exercising, or playing with my children instead.

Karl Marx once called religion 'the opium of the people'. Nowadays, social media likes and responses are opium, for they trigger the brain chemistry that creates positive feedback loops. This sucks us further into machine drift (where the algorithms are the boss and we aren't actively making a choice anymore – we have just drifted off into a sleep-like state, not really aware or in control). The more time we spend feeding the apps and training the algorithms, the better they are at showing us what we want to see. It's confirmation bias in action, fed by our own desire to jump down the rabbit hole into oblivion.

Your data isn't yours

Imagine this: a person with a clipboard walks up to you in the street and starts running through a list of questions without even looking you in the eye. What did you have for lunch? Can you show me a photo? How good are you at dancing to this song? What was your first pet's name? How old is your mother? Where were you living last year? How many times did you eat fast food this week? How many steps did you take last Wednesday?

Would you stand there and give this stranger all that information for free, or would you tell them where to go and walk away? How rude of someone to invade your privacy like this! Who was paying them to ask you all those questions? Why did they want to know all that information? What were they going to do with it all?

It seems outrageous, yet this is exactly the type of information being captured by the robots and AI in your life all the time. Do you know which apps on your phone have access to the microphone or location data? How about your friends' phones and their apps? You're not just an individual – you're part of a huge information ecosystem. Even depersonalised information that mapping apps and the like may collect can actually be used to work out who you are and where you are going.

We pour our hearts and minds out to these robots and AI, and give our information away for free to big tech companies, insurance companies and banks. Do we actually recognise how much time and effort we are giving to the likes of Mark Zuckerberg? What is your hourly rate? Have you considered that you should actually be paid for posting to your social media pages because you are generating so much value to the companies that run them?

Despite there being rules about data privacy, not many countries actually allow you to personally own all the data that is collected about you. (Estonia is a rare exception.) The rate at which technology is evolving has destroyed any concept of legislatively controlled progress; the tech leads the law, and the law increasingly lags further behind. Laws are often passed in haste to try to keep up, and none have been particularly celebrated. Australians will recall the 2018 pushback against government-owned digital medical records – the 'My Health Record' debate. The crisis erupted when the government announced it would create a shareable national electronic health record for each Australian citizen – unless they opted out.

To be honest, I didn't understand the reasoning behind people's reservations. We hand over so much information on social media or through apps that I couldn't see why the government having access to my personal health records would be any different. I am also a big believer in public health research, and I know a number of academics working on rare or difficult diseases who would be greatly assisted by having access to depersonalised information around those diseases. It is not offensive to me in the slightest that academics and scientists would have access to blind data after their ethics applications were approved. It makes me wonder: would people be less offended by the idea of government-funded data collection if they understood just how much they were giving away randomly through their devices?

AI can make you feel things

As a kid in the 1990s, I was the proud owner of a cute digital toy called Tamagotchi. Made in Japan, Tamagotchi was a digital pet that you could look after, feed and play with. Failure to complete these tasks could result in the 'death' of the digital pet. I am sure there are thousands of apps these days that mimic that robot. It was an interesting experience for a lot of us tweens at the time; we all killed at least a few Tamagotchis. Some kids even purposefully neglected or tortured theirs, which was shocking at the time, and as a parent now I would be seriously worried if my kid did something like that. But digital pets don't have feelings, do they? *Do they?*

Humans are great at anthropomorphising robots and AI, but there's a significant ethical issue with the concept of bestowing AI with human emotional capability. Studies have shown that consumers prefer interacting with AI-based customer chat systems to speaking to a real human at a call centre, even if it is just to save time. AI can be used to help match the personality type of call centre operatives to the people dialling in for assistance. Will AI totally replace call centre workers? Possibly. Will we drift towards tolerance of the errors and training issues around AI interfaces? Highly likely. Will we give away more data and information about ourselves to help train these algorithms? Definitely.

What does this mean for the future of free choice and free will? In the movie *Minority Report*, the lead character receives targeted advertisements based on the biometric data in his eyes. This technology exists right now in some parts of the world, except it isn't your eyes that are being targeted but the whole of your face. I use my face to unlock my iPhone and to pay for items using Apple Pay, as well as to access my banking apps and other secure items. You can use your face to pay at vending machines and in supermarkets. AI watches what you put in your basket, and ads for particular products will appear to you in various forms based on your purchases. You can go to work, and AI will measure your temperature and check your ID at the same time.

AI can even detect and respond to our emotions. If we suddenly start adding biometrics to the click-and-swipe analytics that are feeding the algorithms, are we going to start feeding the demons? Imagine if someone is upset or unhappy – will the algorithms show them content to cheer them up, or push them further towards deeper sadness and even depression by providing recommendations that induce gloom?

In the novel *Cloud Atlas* by David Mitchell, Sonmi-451 is a fabricant waitress at a restaurant that is always filled with customers. In her world there is a social expectation that humans spend all of their income for the good of society. They are not allowed to accumulate wealth, and there are spend targets that consumers have to reach to meet community expectations. These spend targets are reminiscent of the unspoken obligations in Western society to consume, and the societal expectations (and norms) around fast fashion and upgradable tech. The AI in your loyalty card databases and data amalgamation service companies can already work with complexity and decide how to maximise your spending: targeted adverts, bespoke content, personalised discounts... sound familiar?

Cosmetic companies and media create expectations about how we should look (and feel), and yet we are trapped in an irony of catfishing whereby the filters on social media are so extreme (and so normalised) that no-one really knows what other people look like anymore. This has led to anti-filter influencers countering the filters and the fakery across social media, as well as social media apps such as Clubhouse and Twitter Spaces, which are audio-based and have no imagery involved at all. Clickbait and chumbox are so prevalent across the internet and social media now that there is an active backlash, with adblockers and satirical websites taking a stance against the pervasiveness of this type of advertising.

Explainable AI and machine learning

There has been some really interesting work done by human rights commissions and leaders around the world into setting precedents so we are prepared and on the front foot when it comes to how AI is going to be taking, using and making decisions with our data in the near future. There will be legal backlash against the opaque algorithms and data that companies and governments are using to make assumptions about us as communities and individuals.

The future is more personalised but less personal, more tailored but less bespoke. Our data creates chimeric doppelgängers of us, rather than digital avatars. This means that fintech, proptech, healthtech and social media companies are making assumptions about us and creating models to fill in the gaps and cracks. These are the algorithms that are testing you all the time to see what you like and what you don't like, but they also target you based on what your friends and family like, too. It's your network feeding the software with data; the software then tests what it thinks you might like. Feel like a lab rat yet?

How can we find out how these assumptions are being made? The answer is to smash open the 'black boxes' of the systems we are using (and feeding). To turn a black box of secrecy into a glass box of openness, we need to employ a methodology called 'explainable AI' (XAI). XAI is already kicking off in the USA, where health insurance companies have been required, through case law, to reveal their data management and analysis algorithms and processes. In Australia, the government exposed COVIDSafe (an app designed to help identify people exposed to COVID-19) to the coding community, which then found glaring errors – for example, that not all states and territories were included in the code. This ability to shine light into the darkest corners of assumption and bias is incredibly liberating for those of us with deep concerns about trusting opaque systems, although it does clash with the idea of commercial in confidence and IP-related business valuation.

We can also use machines to help us decipher machines. You have probably heard the term 'machine learning' (ML). ML is a form of AI, and it is important to understand how it works. AI is effectively a series of equations, algorithms, code and other data-handling mechanisms and mathematics that allows us to decipher and process massive data sets to answer set questions and solve problems. ML, on the other hand, involves computers using data to discover how they can perform tasks without being explicitly programmed to do so. What many people think of when they hear 'AI' – the ability of computers to think for themselves and program things without needing humans to enter the code – is actually ML.

Being able to see the code that is controlling us and challenge the assumptions and errors we may find in it allows us to understand the processes and the necessary lies that are used to glue systems like this together. Can we all read code? No. However, we have AI that can read code and decipher it for us, and we have a new generation of lawyers who will be looking to find faults that cause injury or inconvenience for us and their clients. A whole new industry is being born in this space, and we likely don't have enough people currently trained in coding languages to meet demand.

Get back in the driver's seat

So, what should you be most aware of in the world of machine drift, algorithm vortices, confusion about consent, and computers making arbitrary yes-or-no decisions? What can you do to claw back some control of your own mind? The solutions can range from reading the terms and conditions on the apps that you download (yawn) to getting educated in AI, ML, coding, and new and emerging legal issues. (Contact your local TAFE or university – there are so many courses out there.) Follow the work of the Australian Human Rights Commission, and also Ed Santow at University of Technology Sydney.

I also suggest having a 'dumb' phone (yes, you can still buy an old-school Nokia), or translate your mobile device into a dumber version by moving apps out of plain sight, disabling notifications, switching off location settings, removing apps entirely, or having a lockable container that runs on a timer that you can pop your phone into.

I used to get frustrated that my mum didn't always have her mobile phone on when I tried to call her. Now I can see the logic.

2

The fifth industrial revolution

The future of economics

'A human-centered society that balances economic advancement with the resolution of social problems by a system that highly integrates cyberspace and physical space.'
– Definition of Society 5.0, Japanese Cabinet Office website

The record scratches to a halt. Hang on a second – you're saying we are in the fifth industrial revolution now? What happened to the fourth – the internet of things, connectivity, sensors, smart cities and digital twins? Are we past that already?

Let's check backwards before we move forwards. These are the industrial revolutions we have already had:

- 1.0 was around the 1780s, when steam power was really cranking and making machinery move.
- 2.0 was around the 1870s, when mass manufacturing and division of labour was applied.
- 3.0 started around the late 1960s and was the time of computing.
- 4.0 was around the year 2000, when sensors started getting smaller and more connected. It was dubbed the 'internet of things'.

Industry 5.0 will be about the robotics we put inside ourselves – bionic augmentation and the 'internet of bodies'. It will be powered by purpose, not just profit. It is our chance to take everything we've learned from all the technologies we currently have and make tech work for the good of humanity and society. Technology exists *for* us. Unless we start owning this conversation, we're at the top of a slippery slope that leads to being owned by technology and the people who create it.

We now have business and investing models that claim to be better for the environment, but there have been no real changes to laws – such as the *Corporations Act 2001* in Australia – to incentivise board directors to challenge the current shareholder-return, profit-driven linear business models. This means we must be early adopters of future-focused business models and lead the change. We must create businesses (and jobs) around this new economic approach, and build sovereign capability and supply chains.

People talk about environmental, social and governance (ESG) investing as a platform for ethically driven business growth and sustainable long-term investor returns. This has been the modus operandi of many small businesses for many decades – it's not new. However, it seems environmental impact and social impact (covering issues such as modern slavery) are only talked about on the fringes and are not part of the DNA of large businesses. Our retirement funds are invested in some of the businesses that are seemingly the slowest to change. Is this what we want? If not, what can be done to incentivise those lagging organisations to get up to speed?

The circular economy and 'polluter pays'

The 'circular economy' is a mighty buzz phrase. It's a hugely optimistic approach aligned with Solow's economic model that I talked about in the introduction, and is often linked to sustainable development.

It's where products and materials remain in use (rather than going to landfill), waste and pollution are removed, and planet Earth is regenerated. It is based increasingly on renewable energy and materials, and accelerated by digital innovation. Rather than only being concerned with reducing the ecological and environmental impact of industry, it aims to transform the economy to be regenerative and create new work practices and culture.

Is the circular economy out of reach with current technologies, given our planet's finite resources and growing human population? Have we passed the tipping point of sustainable consumption? My answer to these questions is that we don't know: we don't have the data or evidence yet. My heart says we have to try for a circular economy, though – we can't keep going the way we have been. To do so would mean the end of society as we know it within my children's lifetimes. The old adage remains true, and it is worth repeating here: 'We don't inherit the planet from our parents, we borrow it from our children.'

Some might argue that this planet's resources are not the only resources we are going to be using this century. We already have companies preparing to mine the moon and asteroids; off-planet resources have stirred the entrepreneurs in Silicon Valley to create investment opportunities and generate start-up appetite. The Japanese Aerospace Exploration Agency (JAXA) has landed a drilling sampler on an asteroid, taken a sample, and landed it back on Earth to analyse the iron ore contents. NASA's Artemis project is headed to the moon this decade, which is causing a public resurgence in love for space missions. However, there are no formalised global treaties or trade agreements based on the resources 'off world'. And, really, can we stretch the resources argument that far to insist on a consumerist approach?

Resources on Earth remain finite and increasingly contested, especially geopolitically. The old favourite phrase 'reduce, reuse, recycle' has really only perpetuated processes such as plastic production, as the recycling element seems to give people licence to just keep using it. Ironically, the biggest problem with plastic is that it's very good at its job. Plastic persists; it doesn't decompose. This is causing hell in marine environments as plastics break down into smaller and smaller pieces, entering the food chain and killing wildlife, and even ending up in us. On average, experts estimate that each of us eats a credit card's worth of plastic (approximately 6 grams) in the form of microplastics each week. This amount of plastic is worrying enough, but once it becomes nano-sized, what is it going to do to our health? We don't actually know, but scientists are predicting it won't be good.

There are lots of plastic remnants in our agricultural soils from things such as fertiliser pellets. Plastic disrupts the guts of micro-organisms such as nematodes and causes an increase in carbon released from the soil. One extra thing we don't need in the climate emergency is another nonpoint source of large-scale carbon emissions. Have plastics been a distraction from the climate emergency, or are they inextricably linked? I believe the latter.

We need to deal with the source of the issue that created this plastic monster: the business models allowed to thrive under our legislative framework that don't consider environmental protection to be as important as fiduciary duty and solvent trading. The producer has not been paying. There should be a financial cost for making profit at the planet's expense in an unsustainable way. The 'polluter pays' principle has never really been taken seriously as part of an economic lifecycle analysis. This could be about to change.

'The way we've always done it'

Doing business with a 'this is the way we have always done it' mentality won't work going forwards. The boardrooms of the world are concerned with risk and financial solvency, but most do not have innovation subcommittees. A recent Australian Institute of Company Directors study suggested that fewer than 3 per cent of company directors have a STEM background. In an increasingly technological world, this is a massive problem. If people at the top level of corporate strategy can't engage properly with STEM and the new digital economies, what does this mean for the future of the world's stock markets? We need to make sure board diversity reflects what the future is going to look like, not what the world looked like 20 years ago. Diversity indices around gender are still subpar globally, but when you also consider STEM knowledge, this uncovers a huge opportunity.

Conservative, slow approaches to change will not survive the advent of exponential technologies and Industry 5.0 – customers will march with their feet (and their money). I have already moved my banking and superannuation to organisations that fit with my worldview and morals, and I am not the only one. Industry 5.0 will see a rise in 'profit with purpose' business models, as well as third-party accreditation schemes and expectations around industry best practice. Associations and industry bodies – as well as standards bodies, including the International Organization for Standardization (ISO) – are creating aspirational standards schemes designed to encourage organisations to consider purpose as well as profit. It seems that if you build targets in an industry, people will aspire to meet them.

It has been rare in the recent history of the corporate world to see more than just one or two companies take the lead; the rest are followers. We have seen this with carbon footprint calculators in annual reports, modern slavery law, and post-pandemic rushes to sovereign supply chain management.

Things happen fast in Industry 5.0. This means that regulation and laws lag behind, and it's difficult to have them approved ahead of widespread adoption. The brewing legal and regulatory fight between Uber and some governments around the world never eventuated because the Uber delivery boat had sailed. Uber got what they wanted, which was to operate unhindered. There was backlash against

Uber's expansion into underprepared markets such as Australia; for example, where I worked in 2015, we were told not to use Uber for work purposes, since it was believed the Uber drivers – and therefore the travel – wouldn't be insured.

Ultimately, though, companies like Uber don't actually want drivers – they want driverless vehicles. Drivers are the least reliable part of the operating process, and some software service provision apps are built for a time when humans are no longer part of the front end of the business delivery team. In 10 or 20 years, we may not even own our own cars – maybe we'll rent them when needed, or hire driverless vehicles.

This makes me wonder why we're not creating smart cities and designing new buildings that take the key trends and technological capabilities of Industry 5.0 into account. What will the car parks being built in 2022 actually be used for in 2035? Can they be built with flexibility in mind? Could they help house the homeless and vulnerable in our cities, perhaps? Some charities already do this in large car parks in city centres, such as the Australian not-for-profit Beddown. Temporary beds and infrastructure are brought in after hours and the spaces are transformed into safe, private sleeping areas. Charities such as Orange Sky Australia have been growing their laundry services to homeless friends across the country and are well funded and supported operations. They use spaces such as car parks to offer free laundry services from the back of a built-for-purpose van. These types of charities have many generous corporate sponsors and supporters.

The future of work

The future of work and business are areas of great interest in the futurist sphere. What will the office of 2025 actually look like? Some imagine offices will be divided into smaller, personalised rooms, with more fresh air and natural ventilation, less hot-desking and fewer open-plan spaces. COVID-19 and Industry 5.0 may be the death knell of the open-plan office – which is great news for me, because I have a voice for theatre and used to feel totally hamstrung by large, open workspaces. Architects and designers should be considering how buildings can be designed to fight obesity (such as through better access to green spaces, nice walkways and cycle tracks) and create positive spaces that promote good mental health (such as through providing art spaces, or social, communal space), as well as collaborating to plan healthy buildings fit for the next pandemic (for example, ones with windows that actually open, providing better airflow). Former Lord Mayor of Sydney Lucy Turnbull has also highlighted the need to design liveable cities to be safer for women.

We now have an emboldened attitude towards fully flexible working conditions. Just before the 2020 lockdowns in Australia, I had a friend working for a large bank who begged to work from home on Fridays – she had young children, and the time saved commuting and organising child care would greatly benefit her. She begged and begged for weeks with no response – and then suddenly the pandemic was upon us, and she was given permission to work from home, no problems. It took the pandemic to force management to accept that they could actually trust people to work from home, and in fact to be even more productive. One thing for sure is that office work is going to undergo a hysteretic effect: it will never be what it was now that we know it can be different.

Industry 5.0 could foster flexible business practices that help women and working parents all over the world live happier and healthier lives. We know, on average, women around the world retire with much less in the pension pot than men (some 30 to 40 per cent less according to the World Economic Forum). This is now causing ripple effects that are tipping older women into poverty. With ever-improving healthcare increasing life expectancies, our pensions will need to stretch further. One possible solution is a universal basic income (UBI) or some other form of improved financial support, which would allow women to take career breaks to raise children without being disadvantaged in retirement. Imagine the difference it would make if women were properly financially supported when they needed it most; if rewards based on key performance indicators (KPIs) were not affected by terms of parental, family or carer's leave; if superannuation was paid during periods of unpaid leave; or if other types of leave, such as menstrual leave and domestic and family violence leave, were standardised.

During the COVID-19 pandemic, the Australian government allowed people to access two lots of $10,000 from their superannuation. Many women who were trying to escape violence took advantage of this and were able to move out and away from danger. However, the consequence for those women – who are already disadvantaged in terms of their superannuation – is that they now have even less in their future savings pot. What if those women were given grants instead to save their lives and those of their children?

Expectations around work-life balance are being challenged in our post-pandemic world, and there are some radical changes coming to our standard working week. In the past, in many companies and cultures, overtime had become standardised, and there was competition among colleagues to do the most unpaid work. In Japan, a term was even created for the young people who literally worked themselves to death: *karoshi*. But it seems that the tide may have turned, and the very notion of unpaid overtime is now seen as toxic in many workplace cultures. This change of opinion was always coming, but the pandemic accelerated conversations about not just what work-life balance *could* be but what it *should* be.

In some countries, such as Sweden, people are hired for five days per week but only work four, with the fifth day spent relaxing their neurons and their bodies, and getting inspiration for the next week. New Zealand Prime Minister Jacinda Ardern is also a proponent of the four-day work week. Imagine if we could all work a four-day week, and then spend one day learning about new technologies at TAFE or university? This would add value to the individual, the company and the country, not to mention supporting the struggling education sector.

Some Scandinavian companies have been known to have desks automatically lift up so their users can't reach them after 5 p.m.! In many other countries, people are expected to do a five-day fortnight in the office and work the remaining days remotely from wherever they like. In London, people who are working in the office on a Tuesday, Wednesday and Thursday have adopted the nickname 'TWaTs'.

The idea of trust in the workplace, and KPI-driven performance rather than just hours on the clock, has been an ideal for a long time. The pandemic finally lit a fire under some managers to actually trust their employees to perform while working remotely, though the cynic in me wonders whether this will last unless there is a legal reason for organisations to maintain this.

Cybernetics – the way forward

Cybernetics is the cornerstone of Industry 5.0. It is where the engineers of the future are created, and where the lessons from past opportunities can be applied to create real and lasting change.

Cybernetics can be described as the intersection of humans, technology and the environment. Some academics apply complex and deep technical language to it, but the truth is that we all live, work, play and travel on cybernetic systems. A building is a cybernetic system; a thousand-year-old fish trap is a cybernetic system. It is effectively

Industry 5.0 in action, and reaction. The 'why' of a cybernetic system is to provide solutions to problems in a balanced way, in which inputs, outputs and impacts are all well understood. It is how we define and understand problems and then find a solution through systems-led thinking.

Systems-led thinking acknowledges that digital or technological systems, such as AI, ML or drones, have a job to do. That job needs to be taken into consideration as part of an overall system – an eco-system, if you will – of different nodes doing different jobs, with different levels of interactivity and feedback. This means we need to approach interactions as though they are networked – that is, every-thing is connected.

Imagine that everything we do is an action, and those actions affect things over space and time, sometimes in ways we never know; in physics, this is known as 'string theory'. A technological network is more dynamic and flexible than a food web in nature because there are 'rules' that govern nature – such as that predators are never predated by their prey (or are they?). These rules don't exist in technology. In a technological system, the sheepdog could shepherd the sheep, but the sheep could also shepherd the sheepdog. It may sound counterintuitive, but the rules in automated technologies and machine learning are so flexible that there are really no rules. The environment usually selects the utility of the human and the technology in a perfect trifecta.

This takes me back to my PhD in microbial ecology and the early works of Dutch microbiologist Martinus Beijerinck, who stated 'everything is everywhere, the environment selects'. Let's apply Beijerinck's idea to cybernetics and the environment experienced by humans and technology. It is then a three-way model of averages trying to find the best fit to solve a challenge or problem. This is where things get exciting – when the evolution of society and its tolerance and trust of new and emerging technologies meets with the hungry engineers and technologists who are designing solutions, iterating fast, failing early and evolving quickly.

It works best to understand a problem before trying to design solutions. For example, I first interacted with drone technology when I was looking for a better way to monitor turtle nests and tracks offshore on sandy islands. None of the traditional methods were quite fit for purpose, nothing had been created yet as a solution, and the long-range environmental survey drones were right in the Goldilocks zone of operational and data requirements – they had just the right capability for just the right price. Serendipitous. It refreshed my tired, PhD-trained brain into discovering novel solutions that created new lines of business, real projects and potential research collaborations with universities and start-ups.

However, this approach to activating innovation and radical change depends on having a culture of acceptance and interest in doing new things and doing things differently. Isaac Newton's third law of motion states that for every action there is an equal and opposite reaction, and this is true of innovation and change: there is a brick wall of inertia when it comes to implementing new methodologies, especially if you weren't the supposed 'boss' who thought of it first.

Workplace and boardroom culture aside, there is something to be said for recognising the fact that cybernetics is not new; it is just applied to novel technologies and concepts. Steve Jobs had it right and was describing cybernetics without maybe realising it when he said, 'It is in Apple's DNA that technology alone is not enough – it's technology married with liberal arts, married with the humanities, that yields us the results that make our heart sing'. It is exciting, then, to see progressive arts companies such as the Australasian Dance Collective working on pieces that are live-choreographed by AI, or that integrate with the likes of drones, robots and exoskeletons. Dance, and therefore human expression, married with cyber-physical systems: a perfect date night.

So, what of these new cyberneticists, these engineers of the future? Where do they sit in the current business environment? The answer, it would seem, is 'waiting in the wings'. Systems-led thinking and multi-dimensional opportunities have been stalled by the panic of the COVID-19 pandemic, which is counterproductive – if we were more open to new ways of thinking, we may have been better prepared to respond to and manage the pandemic.

This comes back to the importance of teaching STEM skills and ways of approaching problems to those who will lead our planet into the future. Systems-led thinking, which naturally evolves into project-based learning in diverse teams, is the best way to curate excitement around technology. In Australia, we have a Digital Technologies curriculum that is not about Information and Communications Technology (ICT) – it's about problem-solving using technology. These are the skills and approaches we need for dealing with the gnarly issues we are going to face in the next 10 years. We must be creating school graduates who have problem-solving skills and experience working with diverse teams.

Most Nobel Prize winners these days are teams. Any of us who have worked in teams will recognise that solving problems and innovating is a team effort, and so I am personally really happy to see wider teams celebrated – because no futurist is an island.

Get back in the driver's seat

So, what is coming next? It seems we have worked out that constantly reacting to problems is a very expensive way to do things. Proper planning and preparation, as well as more prescient governance mechanisms, are needed in corporate culture as well as government strategy. To make frameworks, language and strategy more forward-thinking requires conversations about the capability and scalability of new technologies, and the work of cybernetics, as we pass through the next industrial revolution.

Check out Distinguished Professor Genevieve Bell and the work she has started at the Australian National University (ANU) School of Cybernetics. Professor Bell's TED Talk on scaling AI ethically has had more than two million views, and I highly recommend it. Her ABC Boyer Lectures are still the best I've heard, and her team is an internationally recognised and award-winning motley crew working on research and education projects, including creating a Masters of Cybernetics program.

3

Little data is big data

The future of personalisation

'All those moments will be lost in time, like tears in rain.'
– Roy Batty (portrayed by Rutger Hauer) in
Ridley Scott's *Blade Runner*

Ever had a relationship with someone who seemed to know you better than you know yourself? Maybe a parent, friend, lover or therapist? Well, now robots, sensors, cameras and apps on your phone are that someone. They collect a huge amount of data and micro-information about your little quirks and idiosyncrasies – including the way you use your store loyalty cards and credit cards, the way you walk through a shopping centre, and even the way you type into a keyboard. It's all being measured and archived.

How much do you know about the data that's collected about you and how it is used? Where does all the data end up, and is it safe? Do you know how to protect yourself from having your data hacked? How could you and your behaviours be analysed and predicted? Does your location data show you went for drive-through fast food three times last week instead of going to the gym? The drops of data gather to form your personal data ocean, and may well converge to create such a thing as a personalised digital twin.

People my age (I'm in my forties) never studied data management in school, let alone learned about smart devices or social media. My generation is the last to bridge two worlds: a world that understands the relationship between a pencil and a cassette tape, and a world in which almost everyone was born with a smart device in their hand and doesn't know or care what a cassette tape is. Did you see the video that went viral of a man who challenged a couple of 17-year-olds to dial a phone number using an old analog phone in four minutes? They couldn't do it. One of my peers told me of a friend whose pre-teen child found a floppy disk in the home office desk drawer and asked her why she had 3D-printed the 'save' icon.

The reason I find these amusing examples so worrying is that they demonstrate the translation issues that still exist between the digital and analog worlds, and these affect business, risk, finance and regulation. Data literacy is a socio-economic issue as well as a generational one.

Using data to fight crime

Location data, also known as geospatial information, indicates where you – or, rather, your devices – are in space and time. Mobile phones have been used to triangulate lost victims in search-and-rescue operations, and network data has been used to solve crimes – such as the well-publicised 2002 murder of young girls Holly Wells and Jessica Chapman in the UK, when Jessica's mobile phone pinged a local tower when the murderer switched it off. New technologies can now pinpoint your actual location to within 5 centimetres when you make an emergency call, and you can still find a new iPhone after it has lost power or been switched off.

Forensic research has always had a close relationship with new and emerging technologies. Great examples include DNA fingerprinting and ground-penetrating radar that can scan for buried bodies.

This application-driven aspect of new and emerging science gets a lot of attention and funding. I think that is – at least in part – due to the ability to prove the return on investment (ROI) for investors and shareholders.

Many people willingly send their DNA to companies that locate their heritage and ancestry information. I have done this, and it was good fun to see where my genetics have derived from; I also ticked the box to allow the company to use my DNA for research purposes (I am a scientist, after all). The ethics of using DNA database submissions and samples from innocent people to track and identify criminals is still up for debate, but the data is already being used in this way. Because some 20 people uploaded their DNA samples to find their family tree using a company in California called GEDmatch, police were able to identify and successfully prosecute the Golden State Killer in 2018, after four decades of work. I am comfortable with the idea that my DNA – if it were used for research or solving crimes – may help make the world a better place.

DNA fingerprinting isn't the watertight technology it used to be, and genetic identity manipulation could already be used by organised criminals to avoid detection. When technology is demonstrated to be fallible, scientists design and test newer technologies. One new idea is 'active biometrics': instead of just the words, numbers and symbols we currently use to create passwords, the way you personally type that password is also unique, like a password within the password.

Data and diagnosis

Biometric analysis can also be used to detect Parkinson's disease, be it through wearable accelerometers or through video-based AI software for personal identification and gait analysis. This field of research can also identify children with autism spectrum disorder through their foot placement. Such swift and automatic disease and neurodiversity diagnoses could lead to faster interventions, treatments and support for the people affected. However, it could also lead to issues with health insurance or financial risk rankings. Imagine if your bank knew you had a neurodegenerative disease before you did – would they tell you, or would they just blacklist you for credit?

Your shopping habits can also let the cat out of the bag when it comes to personal health matters. A now infamous example of this is the 2011 outing of a pregnant teenager in the USA. Target's algorithms had figured out the teen was possibly pregnant via the data captured from her loyalty card and sent her coupons for baby items, such as cribs and cute baby clothes. Her father was furious that his daughter was being targeted with such advertising and called the manager to complain. By the time the manager investigated and called the father back a few days later, the teen had come clean, and the father apologised to the manager. The fiasco caused many retailers to rethink their targeted advertising, figuring out ways to promote targeted products while avoiding their ethics being called into question. They might work out you're likely to be pregnant, but their marketing will now mix the pregnancy deals with non-pregnancy deals, such as for gardening equipment, so as to make it less obvious that they are delivering you bespoke, personalised content.

Since the COVID-19 pandemic, it has been interesting to see the proliferation of wearable technology such as the Apple Watch including options like oxygen monitoring. Personalised medicine via wearable technology can be extremely beneficial, especially when dealing with something life-threatening such as going through chemotherapy. A closed-loop health service in a remote community could, for example, have a set-up like the following: a patient undergoing chemotherapy has a wearable device measuring skin temperature, heart rate, oxygen saturation and so on. When their pulse starts to quicken and skin temperature starts to rise, their device sets off an alarm at a designated doctor's office, leading to a video call between the doctor and the patient. Then, as they are speaking, the doctor diagnoses a potentially deadly infection, triggering an order of antibiotics from a local pharmacy or community medical chest. While the doctor and the patient are still talking, a drone is deployed to deliver the antibiotics to the patient, so they can start the course immediately with real-time monitoring and follow-up by the doctor (and the technology) over the next crucial hours.

All the technology required to do this exists, but it is siloed – we lack the systems-led approach and the funding for such a thing. As in most medical or health scenarios, there are potentially many great solutions that would pay for themselves in lives saved, but there is little upfront investment in such systems.

Data's dark side

We all have stories of social media advertisers 'reading our minds'. My friends and I like to play with the algorithms sometimes. We know that apps such as Instagram have their microphones switched on and are therefore potentially always listening, so we say the same product name over and over again, and then that product miraculously appears in the advertisements on our apps. Data is even being harvested from your friends and family to better target you for sales. You might be targeted for certain brands of cars or cream cheese not because of what your phone heard you saying, but what your friend's phone heard them saying.

Jetson is an equally interesting and nightmare-inducing piece of technology created in the USA. It uses lasers and a technique called 'vibrometry' to pick up your heartbeat, even through your clothing and from up to some 200 metres away. If this isn't enough to give you the heebie-jeebies then add to it the fact that the rhythm of your heartbeat is unique to you, just like your fingerprints, irises and DNA. This means that you can be identified by Jetson in a crowd, and it is much more accurate than gait analysis or even long-range facial recognition. So, using Jetson for finding a wanted terrorist? Great! Using it in shopping malls or streets for targeted advertising? That's less likely to pass the water-cooler test. What if your phone could pick up your pulse from your hand or ear?

Worryingly, there are even algorithms that can predict your voting habits, and this can then be used to target advertising to try to sway which way you'll vote. A commonly referenced example of this is, of course, the Cambridge Analytica scandal, in which Cambridge Analytica collected some 50 million Facebook users' data without their consent for the purposes of political advertising. The victims had completed personality tests – for which they were paid a few dollars – via an app, and their answers were linked to their (and their friends') personal data on Facebook, which was used to target political advertising during the 2016 US Presidential Election campaign. I don't think the founders of Facebook ever imagined a social media platform could shape the outcome of a presidential election, and yet here we are.

The legal and ethical questions around data and how it can be handed over (sometimes unwittingly) will provide some of the most interesting and pressing debates of the next decade.

Where does all this data go?

Think about how long you've had your social media accounts. It's potentially up to 15 years or so – 15 years of clicks, likes, comments and shares. All this adds up to a lot of data, not to mention all the personal information that you store in the cloud, as well as the never-ending, self-filling porridge pot that is your email inbox. Then add your WhatsApp, Messenger and iMessage use, and don't forget your accounts on Netflix, Disney+ and other streaming services. The world is working more digitally than ever before, and every minute the bank of data grows exponentially.

We heavily rely on data storage that might not even be physically in the same country as us, and we expect immediate access to all that we need. My mum always kept our baby photos in albums in the dining room cupboard (they were the one thing she would try to rescue if there was a house fire, she always said), but my own babies' photos are all on my interconnected devices and iCloud, and then backed up with a third-party online storage provider. So, effectively, I have six or seven copies of each of the 20,426 photos and 1198 videos on my iPhone. I also have copies on my social media accounts (but I don't show my kids' faces there), and my husband also has his own photos and videos (and copies of these) across social media. We share these images and videos across different communication apps and even bog-standard text messages. What on earth does this mean for the supply-and-demand function of data storage and the legal and financial (and climate) issues that come from this?

Some have calculated that 21 per cent of the world's electricity usage will be consumed by data centres by the year 2030. We aren't even talking about analysis and data processing here. Nearly half the electricity used by data centres is solely for cooling purposes, so it was a wise move for Facebook to establish data centres in Luleå, Sweden, some 70 miles south of the Arctic Circle.

Some data centre companies are already calling themselves 'carbon neutral', or aiming to offset their carbon footprint and use renewable energy to power themselves either on site or by building their own solar farms. Green data storage is in our future, that's for sure; it will be pushed there eventually by ESG investing and a shift in the culture of founders, shareholders and company directors.

A lot of data is collected under contract or business arrangement, which means the collector is expected to keep all the data (in all forms – raw, processed, archived, post-production and reporting) for a defined period of time (say, seven years). This means that not only do you have to label, store, back up, archive, share, transport and handle this data in accordance with any data or information management plan your organisation might require, you also have to guarantee it will be in an auditable form for legal and insurance purposes for a certain number of years after the project has finished. This is getting exceedingly time-consuming and expensive, and the data won't stop coming.

The biggest worry about data is 'data tsunamis' – data that gains momentum and volume through the commingling or exponential growth of platform technologies. Some estimates state that 90 per cent or more of all the data that currently exists across all our global systems was created in the last two years. We need to make sure when we collate, collect, merge and archive data that we do it within standardised metadata frameworks, and we create handlers for the data from AI based in deep machine learning (because it would cost too much to use humans for this). This could be something that quantum computing is applied to in the coming decade, alongside new journalling or registering systems.

The fundamental usefulness of data depends on us being able to interrogate it, and quickly (and with accurate results). Our society has come to expect instant gratification and responses, especially where technology is concerned. AI will be a necessary tool for us to handle the quagmire that could be created by data tsunamis.

Data security

One fundamental premise that must underlie all these digital systems is that they are safe. The issue of cybersecurity has never been more important than it is today. Every single human being on this planet is at the mercy of the cybersecurity systems and processes that are put in place to prevent the hacking, disruption and deletion of important data. This is something that our governments, banking institutions, insurance companies and data storage facilities are constantly concerned with.

Cyber attacks are increasing and are already so numerous that in Australia alone there are some 200 attempts reported every day, but many are blocked and don't get through to even being reported. Next time you have a few minutes to kill, google 'cyber threat maps', and you'll see that the number of attacks is much higher than what gets reported. With one cybercrime reported every 10 minutes in Australia, there need to be more conversations about what cybersecurity means for us as individuals, as well as our workplaces and governments. I have lost count of the number of friends who have had their social media accounts hacked in one way or another. Some forget passwords and have old, defunct email addresses as their main email contact for them, while others have experienced full-blown identity theft. I vividly remember when my email account was hacked 15 years ago; this was before proper internet banking and smart devices, so if it happened now I would be scared that my whole life would collapse like a house of cards and I would be completely compromised financially and reputationally.

Get back in the driver's seat

So, where does all this leave the educated but somewhat bewildered person who is wanting to keep their data safe? The best advice I can give is to switch on two-factor authentication for all your social media and email accounts. This means people can't just steal your accounts through a simple hack such as a password shared accidentally, or a data leak from a trusted organisation (and we have seen a lot of those, such as from Marriott International, Oxfam Australia, MyFitnessPal and Canva). Two-factor authentication is opt in rather than default, and I think that is the wrong way around; it should be part of cyber hygiene and community responsibility. I am getting sick of being targeted by scammers who have managed to access my connections' LinkedIn and Instagram accounts.

You can also visit haveibeenpwned.com, which allows you to check your email and mobile phone details to see if your information has been compromised. The nefarious distribution of personal details to spammers is tech-abuse, and is (shamefully) on the rise.

If you're interested to learn more about technology regulation, I recommend engaging with the thought leaders at the Tech Policy Design Centre at ANU. The report *Tending the tech-ecosystem: who should be the tech-regulator(s)?*, which is accessible on the centre's website, is a great place to start if you want to learn more about the structures and issues guiding tech policy, and the leaders in this field.

Global educators can look to the Australian Digital Technologies curriculum, which contains aspects of cybersecurity education throughout. All Australian school students in Years 7 to 10 (ages 12 to 16) are taught this subject matter, but parents and grandparents might be wise to ask these students for advice. Wherever you live, there will be education specialists in this area (such as ySafe in Australia) – ask if they have online resources and training you can access.

Factory equality and worker solidarity

The future of manufacturing

'First Law: A robot may not injure a human being or, through inaction, allow a human being to come to harm.

Second Law: A robot must obey the orders given it by human beings except where such orders would conflict with the First Law.

Third Law: A robot must protect its own existence as long as such protection does not conflict with the First or Second Law.

Zeroth Law [added later]: A robot may not harm humanity, or, by inaction, allow humanity to come to harm.'
– Isaac Asimov's Three Laws of Robotics

In England in 1779, a man named Ned Ludd was working in a hosiery factory. He was rather perturbed by a newfangled piece of machinery that had been brought in: a stocking frame. So distressed was Ned that he might lose his job to this piece of more efficient equipment that he smashed it to pieces. Within a few decades, change-resistant 'Luddites' wrought a wave of destruction. The new machinery itself was not really the target of people's anger, but destroying it was an easy way to upset the factory bosses when workers were fighting for their rights.

Ned Ludd could not be blamed for this because he didn't actually exist. He was a made-up persona designed to frustrate the government. Ned Ludd was fake news.

And so there still resides the ghost of fake news past, the dear Ned Ludd, in the flat earthers, COVID-deniers, QAnon supporters and anti-vaxxers of today – people who propagate false narratives and commentary so much that the noise becomes the signal. By and large, people have common sense about the pros and cons of new and emerging technologies, but trust levels waver when the media peddles mixed messages and fake news is spread.

I believe that robots and other technologies can allow us to be more human in so many ways, across both our work and personal lives. When robots do what they do best – laborious and repetitive tasks – we have more capability to be creative and social. This is Industry 5.0 in action: making technology work for the good of humanity.

Doing the dirty work

Many of us have worked dull, dirty, dangerous, boring or difficult jobs. It blows my mind to think of how many hours I have wasted typing away into Excel spreadsheets that I was never formally trained to operate. I still remember the day someone showed me what the $ sign did in Excel (it fixes a formula to a cell), and it saved me weeks of time. During my PhD I would hurt my fingers from typing in so many numbers, but at the time I didn't know how to create a macro to automate my work – we just weren't taught these things in the 1990s and 2000s. Some of the other equipment I used during my PhD research was older than me. At the time, there were DNA extraction kits and quantitative PCR light cyclers readily available that other groups got to use, but I still had to do the old-school, el cheapo phenol-chloroform DNA extraction method under a flow hood. I not only gained experience under the fear of burning a hole in my hand or my lungs, but I also gave myself my first-ever migraine: I am over 1.8 metres tall and thus not the size of person flow hoods were designed to be used by, so my PhD stood for 'permanent head damage' in more ways than one.

I can laugh at these traumatic memories now, but it is frustrating to see how science and my research were held back by a lack of access to cutting-edge techniques and technologies that would have given me data and answers more quickly and safely (and likely more accurately). The waste-of-time lab work could have been done in a matter of days instead of months. How much further could I have taken the analysis if I'd had access to the computer I used to write this book, rather than a cuboid remnant of the mid-1990s? Indeed, some of my research team's work could be completely done by robots now, especially the soil sampling and DNA extractions. A colleague's studies in the lab next to mine used a fake stomach to represent gastric digestion (which smelled really bad); someone else I know was dissolving rock samples with acid and harmed themselves; someone else was pregnant during her studies and couldn't progress certain projects for fear of risking her health. It may have been 20 years ago, but the technology was there to do it better way back then. Shame on the people in charge who let the PhD students suffer.

And so, this leads me to one of my current areas of focus in my role at a wonderful, world-leading university with amazing leaders and professors who I am so in awe of. They are the good ones, the forward-thinking ones, the innovative ones, the ones rewriting the rules and the ones reimagining the purpose and opportunity for universities. We literally have a 'Reimagine' project across my faculty. I have seen what 'good' looks like now, and maybe one of the reasons I am writing this book is to answer the question: why is there a gap between capability and action? When we have the technology to do things, we should use them, and use them well.

Reviving flailing industries

Using robotics and automation as a propellant to revive the Australian manufacturing industry is such a no-brainer, I feel like I am stating the glaringly obvious. If we could only see the good that could be done with new and emerging technologies, maybe the general population would have more love for research, and the governments of the world would rally behind it with funding.

One thing that automation, robotisation and technology-based processes bring is a real chance to have an economically viable manufacturing base in countries that mothballed their factories years ago. If you wish to have a manufacturing sector that can compete on a global scale with countries that have lower costs of production and that may not look after their workers to the same standard, then there is just one way to achieve it: automation and robotics-based or augmented manufacture. Robots can work 24/7 (apart from when they need maintenance), they don't take sick days or holidays, and they don't quit and go work somewhere else. We can make our own robots, and make robots that make and maintain other robots.

Robots are great at following instructions and can even iterate to improve efficiency (should you wish for them to have that capability). They are ideal for repetitive tasks. They are also useful from a work-place health and safety perspective – when an accident happens, no human risks being hurt or killed in the process. Most corporations that have drastically improved their health and safety ratings have done so through automation or robotics, not by coming up with new ways of teaching and training humans. Human training takes a lot of time and money; programming machines is faster and cheaper.

Some factories and warehouses now operate fully without major human interactions. Take a warehouse for a business that sells books: drones keep an eye on the codes on the boxes, and remote ground vehicles drive products from selected shelves to packaging robots. Sorting, shelving, archiving, moving, lifting, shifting – all these things can be done without a human in the room. There are a lot of benefits to this – a major one being that you can fit more things in the warehouse.

During the pandemic, 'sovereign manufacturing capability' became a new buzz phrase, suddenly appearing in government strategic plans around the world. The weaknesses of a global manufacturing system and the globalisation of business became very obvious when borders started shutting and transport became very difficult. At the time, a friend of mine who works in the textile industry called around her business partners to tell them to buy any commercial sewing needles they might need for the next year, because there was going to be a shortage of commercial needles and other imported goods required for textile and clothing production. Australia doesn't make sewing needles or cars, has little textile production, and is reliant on the global supply chain for pharmaceuticals. We don't even make jigsaw puzzles or yoga mats (both items that were very popular recently). The shame of it is that we have such a long and diverse manufacturing history here, and we still have people in the workforce with these skills. These people could help build and train robots, and then continually improve the systems and processes for the manufacturing of items in factories.

Developing economies will need support to work with these types of processes, too. As we remove ourselves from the global supply chains that support millions of workers – particularly women, when it comes to textile manufacturing – we need to make sure that those workers don't fall further into poverty.

Where do humans add value, then?

What does this mean for the workforce of the future? It means radical approaches to retraining and diversification of skills, or upskilling.

Removing the human from the loop makes sense for repetitive, backbreaking work, but in most cases there are still parts of the shop floor where humans add real value. In an example I saw recently, fitting robotics into the manufacturing process required new staff to be hired, because roles were created in managing the robots. So, robots aren't going to take our jobs after all, but retraining and upskilling will likely be required. Whole new job titles are being created for the management, maintenance and operational direction of robots – jobs for life.

Robots are very good at doing what they were designed to do, and not much else. This is where AI and robotics are converging on loop systems with very little needed from humans. Imagine applying some of the concepts used in digital health to the kitchen: with smart fridges, online shopping and drone or robotics delivery, and then assistance robotics, in theory you could arrive home to a delicious dinner prepared and served up by 'le chef' robot arms.

The washing machine and vacuum cleaner might seem so commonplace now that we don't even think about them, but what excites my family is the move towards ironing and folding robots (and some do exist). I personally blame ironing for climate change (jokingly), and I think I did too much ironing as a child, so I avoid it at all costs.

The economic value of humans in manufacturing will be in the artistic, authentic, handmade products that will sell for more money because they are artisan and creative: artworks, environmentally friendly beauty products, homegrown veggies and hand-stitched items. The rise of the 'right to repair' movement shows that we're finally rebelling against our current throw-away culture. This desire to fix rather than replace is a throwback to my nanna's post–World War II generation, when the only way was to 'make do and mend'. Resources were precious, and items were invested in to last a long time.

Fashion designers such as Dame Vivienne Westwood base their business models on, 'Buy less, choose well, make it last'. The CEO of Patagonia once asked patrons to stop buying their products unless they actually needed them, and even now you can bring old clothes into their shops to be recycled or upcycled. The fashion industry needs to lead the circular economy, morally and ethically, because it is estimated to be responsible for approximately one third of the world's wastewater, among other environmental impacts, according to the Ellen MacArthur Foundation (which produces some fascinating if not slightly depressing research).

We may trust our manufacturing industry to engage in best practice, but how can we tell whether the cotton in our t-shirts is locally made and not harvested or spun by slaves overseas? The answer is in the smart supply chain – with labelling, tracking and ledger systems. In Australia, we pay a lot to prove that our exports are our exports, especially for products such as beef, wine and honey (yes, there is fake honey), and the main ways we do this are through radio frequency identification tags and even genetic testing to prove provenance – why not apply this to other industries as well?

Exoskeletons

When humans can't completely be replaced and there are still risks that can't be eliminated, the answer is to add the robots to the humans (or add the humans to the robots). A great example of this is the exoskeleton – a kind of armour that a human can wear. Exoskeletons can be very small (even just on one finger) or cover all the major limbs, depending on what you're trying to do. They can be directly plugged into power, work off batteries or work with natural body movement.

Exoskeletons have been used in factories in Japan for a long time already, particularly for heavy lifting and overhead shoulder support.

There is huge potential in other industries such as healthcare, where nurses and care staff are dealing with more and more patients with obesity as well as the pressures of an ageing population. If an average workplace health and safety incident such as a slip, trip, twist or fall (which accounts for the majority of workers compensation incidents in Australia) costs an estimated AU$100,000 or more, the investment of some $5000 to $50,000 per unit (depending on the size and complexity of the exoskeleton) should not be seen as too great a cost in crude financial terms. In reality, the more we use exoskeletons, the cheaper they will become through economy of scale.

Imagine having support for your back and arms so you could easily lift 100-kilogram items. How would this help you move house? How much more of the house could you paint if you didn't fatigue so fast? As a new mother, how many times could I have used an exoskeleton to help me rock my baby to sleep in my arms? This melding of human and machine has a large part to play in bringing back manufacturing to countries such as Australia, the USA and the UK.

Do exoskeletons sound too good to be true? Well, Iron Man they are not. There have been cases of people being injured by the exoskeleton that was supposed to be helping them, which has paused their use in some cutting-edge car manufacturers in the USA. Another issue with exoskeletons is the battery life, because they are very energy intensive, but battery technology is improving. In some cases, this could be overcome with microgeneration, such as building solar or kinetic power generation technologies into the suits. Some technology investors think exoskeletons will never reach full commerciality. I personally hope they are wrong because I want one under my Christmas tree.

It's exciting to think of different business models and the profitability of manufacturing with a different mix of human interventions. Lots of people are investing in and researching how to make automation a positive force for change in this space. They are even asking questions about the personas of the staff who would work in such an environment, and how they can maintain currency through education around the drones, robotics, AI and automation that these systems converge. This is Industry 5.0 in action, and it is a whole new business paradigm.

Here's a different way of thinking: you *deserve* to have robotics and technology supporting you in your life and work. Be it an exoskeleton, a quantitative PCR lightcycler and DNA extraction kit, the latest piece of 3D-printing tech or a drone. We are entitled to work safely, both physically and psychologically. We are – by law – allowed to use the known and available technologies that might be used in our industries or sister industries; soon, legal ramifications will be shown in case law for the managers and directors who refuse to make use of reasonably applicable innovations. You have a right to have a robot if it is proven to make your job safer and isn't an unreasonable ask.

Get back in the driver's seat

It can be tricky to know what's fake news and what you really should be worried about when it comes to new technology. My advice is to use logic and disregard anything that smells like disinformation. Also, if you can, donate to research. Support your alma mater, or research charities or universities that are focusing on your topic of interest. Unrestricted funding provides the greatest research freedom. Attend some public lectures and support the research leaders who are on the cutting edge.

Follow the latest news on robots in the workplace (aka 'cobots'). If you work in health and safety, be sure to get to some tech events and access education offerings. Prevention is better than cure.

5

The last athlete in the metaverse

The future of sport

'In order to protect the integrity of global sports competitions and for the safety of all the participants, the IOC EB [executive board] recommends that International Sports Federations and sports event organisers not invite or allow the participation of Russian and Belarusian athletes and officials in international competitions.'
– IOC statement after the invasion of Ukraine in early 2022

'You didn't win silver; you just lost the gold.' This was a mantra of a colleague of mine some years back, and it remains quite a good indication of the toxic workplace I was in at the time. This mentality that we need to be the best or we're nothing really grates on me; it has no place in a tech-positive, egalitarian future. I think it is one of the reasons we idolise professional sportspeople more than scientists and technologists. We need to celebrate scientists the way we celebrate sports, but that is going to take a major shift in culture.

What is the future of sport, anyway? With arguments about the shoe spikes runners were wearing in the Tokyo 2020 Summer Olympics, and the furore around gender identity and testosterone levels in women's sports, it's clear that a lot will change in the next few years. Interaction between biology and technology, down to the cellular and even subcellular level, is now being researched in labs around the world. Things move really fast, especially when money can be made. And there is a lot of money in sport.

Tech doping and human–robot interaction

In the 1985 James Bond movie *A View to a Kill*, thoroughbred horses are doped mid-race in a way that is undetectable. It turns out they're receiving microchip-controlled drugs developed by a Nazi scientist working for the bad guy. It's funny how I remember this more than any other part of the film. I guess it intrigued me as a teenager – the idea that computers were able to work inside the body.

In another movie, *Innerspace*, US Navy pilot Tuck is shrunk to the size of a red blood cell and placed in a special pod, which then travels around people's bodies (including that of his own girlfriend – and he discovers she is pregnant). Piloted or automated robots working inside our bodies to cure us of diseases or perform diagnostics might once have been the realm of science fiction, but as nanomedicine advances, such ideas have become more and more credible. (We won't be shrinking Dennis Quaid in real life anytime soon, though.)

If we can use technology to target drug delivery live during a competition – maybe even just giving a glucose boost – in a way that is undetectable, anti-doping agencies should be concerned. Add bionics into the mix and we really could go deep down the rabbit hole and never come out. I have a bad feeling that this will make the works of Lance Armstrong and his team of cheaters look like a cheap trick. He was at least using his own bodily fluids when loading himself up with red blood cells. What if he'd had access to nanomedicine-based technologies (operating on the size and scale of DNA) and CRISPR (gene editing technology)?

A colleague of mine has suggested that the technology readiness level – a method developed by NASA of describing the maturity of technologies on a scale of 1 to 9 – of highly personalised tech doping is at the top rung, level 9, and will soon be uncontrollable. They say we therefore have only two options: ditch professional sport altogether, or accept that everyone is tech doping, in which case sport becomes a test of a team's or country's tech doping capability as well as athlete performance.

But I don't think we need to give up and accept this yet. Doping is still so vilified that sponsors and supporters would be likely to just walk away if it became the norm. The best way to proactively support professionals in this area is to ride the first wave of these technologies. If we, as a society, can own and develop them, we can also understand how to detect them.

I do wonder, though, if we'll ever see an honesty system with 'tech' and 'non-tech' Olympics and Paralympics. There are already positive examples of new sports that combine humans and robotics; for example, the Cybathlon, which took place for the first time in 2016, tests both human performance and tech performance. It also places a spotlight on the current and emerging technologies available to increase accessibility for those who need it for day-to-day activities.

There are six major competitions in the Cybathlon, covering a gamut of technologies:

1. Brain-Computer Interface (BCI) Race: pilots with quadriplegia use brain-computer interfaces to control avatars in a computer game.
2. Functional Electrical Stimulation (FES) Bike Race: using electrical stimulation, pilots with paraplegia perform a pedalling movement on a recumbent bicycle.
3. Powered Arm Prosthesis Race: a race for pilots using an arm prosthesis on one or both sides, which has to include the wrist and can be navigated with any kind of control.

4. Powered Leg Prosthesis Race: pilots using a leg prosthesis on one or both sides that includes a knee joint have to perform various movements.
5. Powered Exoskeleton Race: pilots with complete thoracic or lumbar spinal cord injury compete using a wearable, powered exoskeleton, which enables them to walk and master other everyday tasks.
6. Powered Wheelchair Race: pilots with a severe walking disability use a powered wheelchair to overcome obstacles such as stairs or doors.

It is wonderful to see these diverse teams of bionics and robotics researchers working alongside each other to make life better for people with disabilities.

This connectivity between people and technology is becoming more intimate every year. The ways in which these technologies can be used for good may seem obvious, but current technologies are still too expensive to be accessible to all the people who want to use them. As these are exponentially advancing technologies, I am really hoping that the costs start coming down soon so those whose lives would be improved by them can access them, either privately or through government support.

Tech for training and recovery

Athletes don't just wear technology during competitions but also during training and, my goodness, we have come a long way fast. I don't exercise nearly as much as I should (cough) and am desperately trying to find ways to drop my baby weight and COVID kilos. Now that I am in my forties, I also worry about living a long and healthy life to see my children grow up. I *know* that exercise is good for me; I just really struggle with time management, motivation and energy, especially with two kids under five. So, thinking it was the best idea ever, I bought a Fitbit. I mean, who hasn't? Right now, I can't even tell you where mine is since moving house, but I know it is packed away somewhere in one of the boxes in my office, and I should probably find it, because it wasn't cheap. I thought data would act as a motivator for me – that I would compete with myself on cardio minutes and steps walked, and calories eaten. I did for a while, but then it dropped off a cliff into the sea of wishful habits and good intentions.

Although I am far from a professional athlete myself, I can really see the value in accessing constant information around training and activity to help inform peak performance and recovery. Many rugby and soccer players now wear devices that collect convergent health data and GPS location data when training and playing, and this data can be used to determine the tactics that work and the areas where an individual is most at risk of injury. It can also be used mid-match to select and swap out players who are fatigued. The Australian Institute of Sport has an online Athlete Management System, where you can plug in data from different devices and film across multimodal platforms to help inform training regimes.

The gladiatorial nature of some sports will change as lawyers, insurance companies and legislators get involved. Boxing has always been known as a sport where people purposely try to knock each other out, which can lead to head injuries. In sports such as boxing, rugby and Australian Rules football, the frequent (and, until recently, accepted) instances of head and brain damage may become a thing of the past – not through smart headgear but through multi-million-dollar lawsuits and compensation schemes. Small but frequent head injuries, even though they may feel like nothing significant at the time, can cause irreversible brain damage. This is called chronic traumatic encephalopathy (CTE); it's when a protein called 'tau' malfunctions, causing other proteins to misfold, which then sets off a brain-death chain reaction.

Scientists are only just starting to understand this insidious issue, but are able to work on it thanks to the hundreds of military personnel and athletes who have donated their brains to the research teams. CTE can result in serious problems that reveal themselves many years after the last match has been played, such as early-onset dementia, drug addiction and domestic violence. Out of more than one thousand brains that have been donated to date by people who suffered long careers of having their heads knocked about, about two-thirds have been diagnosed with CTE.

So, what does CTE have to do with technology? Well, once we understand what is happening with frequent, low-level head injuries, we can then monitor and measure them. Around the world, we have comprehensive workplace health and safety requirements that have a foundational aim to prevent illness, injury and death at work, which for professional athletes means training sessions and competitive games. There are young athletes today who are undergoing brain MRI scans to detect CTE as early as possible. Depending on what's discovered, it is possible that head injuries will not only kill brain cells but also some of the world's most loved professional sports.

Identifying natural athletes

Another aspect of professional sports in which technology has a significant role to play is in the preselection, scouting and selection phases. Can you imagine being chosen to train in a professional sport because you have a genetic and phenotypic predisposition to greatness in that sport? Like swimmer Michael Phelps's wingspan, or gymnast Simone Biles's jump height, there are some physical traits we can identify early to work out whether you have a natural predisposition for sporting excellence.

This may be one of the outcomes of the data from mass surveillance. AI-based processes can pick up biophysics and use it to predict a disease, but the same AI could also be used to predict physical capabilities. Would you want mass surveillance scouting in your child's school for the next set of medal winners?

Esports

It's not only the traditional sports we grew up with that are being transformed by tech. 'Esports' is the collective term for competitive-level computer gaming. Although organised competitions have long been a part of video-game culture, these were largely between amateurs until the late 2000s, when esports grew – through professional gamers, sponsorship and huge spectatorship – into an industry worth an estimated $1 billion.

It might sound surprising, but computer games are worth more globally than movies and music combined. Gaming doesn't only happen at home these days – professional game tournaments fill entire sports arenas and attract crowds of tens of thousands, particularly in countries such as South Korea. It is on my bucket list to see such a spectacle.

Sixteen-year-old Kyle Giersdorf won some US$3 million in 2019 playing popular online game *Fortnite*. I bet his parents don't discipline him for too much time spent playing computer games.

Esports currently involves human hands on joysticks and control-lers, but could soon involve the brain computer interface (BCI). As far as I know, no gamer has yet had a controller actually embedded into their brain, but the concept of deep, immersive virtual reality (VR) and the potential for new business growth caught the attention of Facebook founder Mark Zuckerberg, and back in 2014 he bought VR start-up Oculus VR for a cool US$2 billion. Zuck's press release quoted him as saying it was a way to 'enable even more useful, entertaining and personal experiences'. VR headsets were originally forecast to be big in the social media, networking and even online dating worlds. But what about sports?

When Formula 1 (F1) driver Lewis Hamilton came on the scene in the 2000s, he said that while he'd never driven a particular circuit in a race, he had already been around it thousands of times in his head: he had watched hours and hours of footage taken from racing cars, so his body would know where the corners were. This was before the deep, immersive VR experiences used in sports training simulators today existed. There are now full-on, suited-up, fully immersive training experiences in which you can run around like you're in another world. You can win that race because you've raced it more times than your competitors; you can score that seemingly impossible goal because you've scored that goal thousands of times before.

This is all great for rich countries with access to these technologies, but will a digital divide increase inequity in the sporting world? VR is not quite doping, but it still provides a huge technical advantage.

A common theme when it comes to new technology is that the law is often lagging, and it's no different for esports. For example, laws around advertising, particularly to children, have not been applied to esports in the same way as traditional sports, because esports is considered a completely different activity, but to me this is completely illogical. It's puzzling how laws that are so sensible and necessary are waylaid. If you can't advertise fast food near my kids' school, then you shouldn't be able to advertise it to them in their own bedrooms.

Sports and the metaverse

The future of sport might include not just humans that are more like robots, but also robots that are more like humans. *Robot Wars* was a very popular TV series in the UK that ran for a few years from the late 1990s to the early 2000s, and was rebooted in the late 2010s, and was pretty good fun. Teams developed robots to fight each other in a pit. I can't say it was very gender diverse, but it wasn't offensive – it was just like a futuristic but much less disturbing version of cockfighting. I find it interesting that the show developed a cult following. For me, anything that gets people interested in science and engineering is a win, and it was fun to see the teams coming together and really loving their robots.

If we were to consider a modern version of *Robot Wars*, would it be based in reality or in the metaverse? The term 'metaverse' first appeared in Neal Stephenson's 1992 dystopian novel *Snow Crash*. The metaverse is a virtual place you've seen before in sci-fi movies such as *The Matrix* or *Ready Player One*. It is a permanent, online, immersive environment that isn't owned by anyone in particular (or is it?). Twitter is a good example of a rudimentary metaverse, where celebrities and trolls are equally as connected to potentially every other user (or bot) on there. Humans, as avatars, can build businesses, meet others and even play sports.

The metaverse could be the world of your dreams or of your nightmares. When I think of the future of the metaverse, I imagine all the people 'asleep' in the movie *Inception* – they were choosing to live in their dreams because the real world was so painful or boring. How terrible would it be if reality faded into greyscale in the face of a hyperreal and gloriously technicolour fake world? Maybe the metaverse would be a place we could upload our brains, so we could in effect 'live on' after our bodies have expired.

There are now entire soccer teams made up of AI-enabled robots competing in technology games such as the RoboCup. In the metaverse, we could also have sports teams made up of AI-enabled avatars. After IBM's supercomputer Deep Blue beat human grandmaster Garry

Kasparov at chess in the 1990s, and then some years later Google's AI platform AlphaGo beat Ke Jie in the more complex game of Go, we are now seeing the rise of AI versus AI. Would you find humans playing esports more interesting than watching two AIs take each other on at chess?

In the 1983 movie *WarGames*, the US military created a computer to work out how to beat the USSR in the event of a nuclear war. The computer almost hits the red button but is stopped. The feisty young protagonist challenges the AI to fight only with itself to see who would win, and as it iterates all the ways in which the game could be played, it realises that no matter how it uses the US nuclear arsenal, everyone loses.

All these wonderful inventions and wearable tech devices will certainly change the way we see sports in the future (literally, in some cases). The key thing it drives in me, though, as a mother of two small boys, is an inherent desire to take my kids to the nearest park to just kick a ball around for a bit, for no reason other than for the pleasure of hanging out with each other.

Get back in the driver's seat

How will we be physically interacting with our children in the next decade? In truth, the way children grow and develop is such a physical thing, my bet is that we will move our families further away from technology in order to maintain genuine connections. We'll diverge from the games console and back to Mother Nature. Being physically active can now be mimicked on a biochemical level by a pill, but metabolic pills don't make your kids giggle insanely the way they do when they launch a kite for the first time, or score a goal against their dad. Playgrounds are not going virtual any time soon.

Remember moderation when introducing technology into the mix, and think about both the physical and mental ramifications of a lack of exercise. Obesity rates in the Western world are soaring. The COVID-19 pandemic lockdowns really made us appreciate being able to get out of the house and exercise. Habits are hard to form, so make sure you give yourself the gift of moving your body as much as you'd like.

As far as professional sports are concerned, enjoy them while they last. Some may change or disappear this decade. And prepare yourself for stories of tech doping from the most surprising and unpredictable places and people.

6

A drone by any other name

The future of aviation

'The busy bee has no time for sorrow.'
– William Blake, *The Marriage of Heaven and Hell*

Every superhero has an origin story, and sometimes that story says more about the present than it does about the past. Back in 1896, Samuel Pierpont Langley was building the first flying machines, called 'aerodromes', and he successfully launched one from a barge on the Potomac River. You could call this the world's first ever drone, because it was uncrewed – it was a scale model of what would be needed to carry people.

Sam was an engineer, a mathematician and a professor of physics and astronomy, and even has something in common with Marilyn Monroe: Marilyn's first job was working to build radioplanes during World War II. These aircraft were used by the US Air Force for target practice before heading over to Europe for D-Day. In Australia, we had the GAF Jindivik as our radio-controlled target drone of choice (developed as a result of a bilateral agreement between Australia and the UK).

But what can drones really do for us outside of defence? The answer: more and more every day. Drones have evolved as an emerging and exponential technology within an increasing ecosystem of technologies. The drone industry isn't really an industry in its own right – drones are now spread across all industries. Your first interaction with a drone might have even been a nuisance down at the beach, or maybe a toy drone that was delivered by Santa (every year the hashtag #DroneCrashmas trends on Twitter). But these days drones are much more than just plastic toys you can buy on the high street. They are made in many different countries and have many and diverse names, types, sizes and configurations: some are the size of a fingernail, while others have wingspans of 25 metres. There are drone bees that pollinate plants in greenhouses, and high-altitude pseudo-satellites (HAPS) that roam at 60,000 feet. Some might even go so far as to call the International Space Station a part-time drone, because it can be controlled remotely when need be.

Drones can create, collect, collate and process data in many ways and at any time of the day or night. They can run, walk, crawl, swim, climb and fly. They can move in four directions and can be used for reconnaissance and counterterrorism; swim underwater to map the ocean floor, or hunt and kill crown-of-thorns starfish; self-navigate in the darkest underground mines or in collapsed buildings; and fly across vast spaces searching for people who are lost, or for rare species of animals or plants. They can even patrol beaches and detect sharks, setting off shark alarms to get people out of the water and harm's way. Every level of government now uses drones or drone-based data collection. There are businesses leading the charge, regulators responding to industry need, and investors backing some incredibly diverse aspects of the drone ecosystem. Australia is leading much of this, especially as the first country in the world to create regulations for drones within model aircraft legislation back in 2002.

Drones do have a terrible name. It makes them sound like mindless zombies. I just can't get past it. It disappoints me that we haven't come up with something better in the 10-plus years I have been working with them. They could be called so many things, not many of them positive or without acronyms that sound aviation- or defence-oriented. The PR machine has let down these amazing platform technologies that are saving lives, helping us protect endangered species and changing how we grow food, among other amazing uses. I accept that it is probably too late to change their name and narrative now, so we are stuck with it. 'A rose by any other name would smell as sweet.' But I'm still not satisfied!

The drone ecosystem

Drones are becoming smarter and faster, and carry technology payloads that are shrinking in size: cameras, microphones, sensors and detectors. The new method of drone deployment is as part of a swarm, and swarming technology is getting better all the time. Swarming – especially multimodal swarming, where a big plane might drop a smaller drone, which drops a smaller boat, and so on – will be a game changer. Not only will drone swarms take on defence roles in the theatre of war, but we will also be able to use them to map and monitor crops, bushfires and oil spills (think drone boats launching aerial platforms), and even look for people lost in the wilderness.

Think of drone swarms as sheepdog and sheep, where they can swap out between roles, or as an ecosystem – like on the African savannah, where the elephants are connected to and working with the hippos, which are in turn working with the antelope. Each has a different role, an ecological niche. It is this type of eco-inspired co-working swarm that is currently at the cutting edge of drones and robotics, and also leading conversations around the ethics of how we operate such 'autonomous' multi-platform technologies using AI.

The biggest kryptonite for drones is the battery technology that powers them – though some of the larger ones have petrol engines, and a few even operate using hydrogen fuel cells or with hybrid engines. The 'new gargoyles' in our cities are potentially on-building or rooftop drone charging stations. Dormant drones placed strategically around our cities just waiting for the call could be used to get defibrillators or insulin pens to people having a medical emergency, and they are faster at reaching patients than an ambulance. We have all seen ambulances struggle to get through bumper-to-bumper traffic, with some drivers not bothering to move out of their way at all. Imagine if, instead, a drone was on its way to a crash site with medicine, blood and a live streaming doctor attached.

A flying ambulance drone such as this is being developed in the USA. There have been a few start-ups in recent years in countries such as Japan, New Zealand and the UK that have been paving the way to personalised aerial transport, with jetpack-like suits you wear or step into, or vertical take-off and landing (V-TOL) drones that can rescue people and then carry them in cage-like structures underneath the engine system. There are a number of global businesses leading the charge for uncrewed aerial personal transport drones to become commonplace. Although the technology may not be available to suit-wearing commuters any time soon, it certainly has wide applications in defence and emergency services today. Would you wear a jet pack, or get into a flying taxi? What if it didn't have a pilot on board? Imagine it's you waiting for that medical help – would you take a flying ambulance if you were in pain and it would get you to the hospital quickly?

We all want help when we need it, but with things that fly or are automated, it is never as easy as it should be to get the required systems up and running. The difficulty with uncrewed, remotely piloted or automated aircraft is that they could kill people if something goes wrong and, as always, the law lags behind: it hasn't been fully tested in a court of law as to whose fault such an accident would be. Uncrewed transport is set to be a game changer for delivery and transport, though. Ultimately, we know that the app-based taxi and food delivery companies don't actually want human beings in the cars, but at the moment the assurance levels for driverless cars or aircraft just isn't at a place where businesses would be able to access insurance. Volvo has taken the lead on this issue, saying it will take full responsibility (with some caveats, of course) for its driverless vehicles – though it doesn't have any flying vehicles yet. Keep an eye on Toyota, which has invested millions of dollars in flying cars in recent years (and have you ever experienced the traffic in Tokyo?). Some time ago, Audi developed a concept car with a drone pack that could be attached to the roof and utilised if you were stuck in traffic. Can you imagine just taking off and flying over all those people stuck there waiting? I think this will be something the police services get to grips with fast – and I'm sure anyone who has been stuck on the M1 between Brisbane and the Gold Coast will follow.

The way drones move around our cities will be controlled by something called remote traffic management (RTM). Singapore was the first city to make real headway in this area of airspace management. Think of it like the Marauder's Map in the *Harry Potter* books, where you can see where everyone else is and how they are moving. Drone corridors will exist in the airspace – not quite like flying traffic lights, but more like shipping lanes in the ocean – using GPS in 4D (time and 3D space). The capability to identify and track every drone in a city's airspace will help people trust drones more; and when you have trust, you have better assurances of how people will behave. When you have decision quality assurance then, as a business, you can look at insurance policies. Once you can insure your drones and their activities, then you have a business model that can make flying delivery drones or healthcare drones a viable proposition for operators and investors.

There are some great success stories of start-ups and entrepreneurs making a good go of it with drone tech, and then there have been the abject failures to meet customer expectations. But despite failures from large companies through to small Kickstarters, it is fair to say that business model innovation in the drone ecosystem and industry is not slowing down any time soon.

Drones and Mother Nature

I am really proud of the world-first drone projects I have created, curated and directed. In one such project, long-range reconnaissance research drones – made in Australia for weather monitoring 20 years or so ago – were used to fly hundreds of kilometres and map offshore islands and marine systems. My initial projects all took place nearly 10 years ago now at the time of writing, but the core long-range turtle-monitoring project still remains ahead of its time. And the imagery, oh my gosh – it would have made Sir David Attenborough want to buy me a beer. We were able to capture footage of Mother Nature in action and, most importantly, without her knowing we were there. Because of the size of the aircraft and the altitude it was flown at, it didn't enter any ecological windows that would cause a predator–prey reaction from the animals we were observing (a lot of drone-based work causes a huge amount of disruption). Even the seabirds didn't notice us operating overhead; it was like we weren't even there.

It is this that any ecologist wants: no sampling bias. It really upsets me when I see people winning drone photography competitions based on endangered species, such as eagles, interacting with them. This is technically non-predatory predation and will have a negative impact on the animals in time, if not immediately, through fatigue, stress, or even cuts and physical damage – either directly from the drone itself or from hitting something while trying to get away from the drone.

In Africa, drones have been used with some success to help protect endangered and protected species, such as rhinoceroses, by providing surveillance of the animals, as well as chasing the poachers. Well-funded and organised professional poachers are now able to use drones to find the animals they want over long distances, so rangers are now engaging anti-drone technology to try to take out the hunter drones. In northern Australia, a project co-funded by Microsoft uses drones to locate feral pig breeding sites in order to inform the professional hunters who go in and take them out. Turtle nesting areas are mapped and then protected using plastic mesh 'lids' that have holes large enough to let the baby turtles out, but are strong enough to stop feral pigs from digging up the nests and eating all the eggs.

Drones have operated in our oceans for decades, from drogues (which are essentially GPS-capable buoys that are passively floating around on ocean currents) to underwater gliders that can be activated via GPS or sensors to seek certain attributes in the water or map seafloor. Slocum glider drones were used during the Deepwater Horizon oil spill in the Gulf of Mexico to identify where the oil spill was in the water column while using on-board processing and special laser sensors to prove the oil was from the leak, not just some old boat engine oil. There are also long-range sailing drones (from a company called Saildrone) that are collecting information and sending it live while traversing the Pacific Ocean. It is wonderful that we can log on and see some of these drones in action – they are operating in places that humans might never go to and providing some amazing datasets that are changing the way we understand our oceans.

Around the world, the rate of coastal erosion has been on the rise, which means a lot of our coasts are now managed or engineered, and sadly in some countries are now more concrete than natural systems. We are using drones and mapping to find new ways to monitor the effects of wave action, and the erosion and deposition this causes. From the severely eroded post-storm beaches north of Sydney to the flooded levees of southern USA, drones provide immediate and high-resolution imagery to assist engineers and coast managers to save houses and roads from being lost to the sea. They are also able to map immediately after storm events to provide information to insurance companies and planning authorities so the right post-event responses can be triggered.

Consider what using drone-based, highly accurate mapping data could mean for the Pacific Islands as they face climate change head-on in the form of rising sea levels, increased strength and frequency of damaging storms, and environmental damage and impacts on natural resources. Small start-up businesses are using drones to provide location and mapping-based spatial data processes to do everything from counting coconuts that are harvested for export, through to counting the roofs lost in storms to trigger relief payments. International aid bodies have provided money for skills and equipment to be provided to local geospatial teams to build local capabilities. After Cyclone Pam bashed Vanuatu, relief flights were taking off from Australia without knowing if the runway was operational at Port Vila because a proper survey had not been able to take place. A drone team would have been able to provide information almost immediately via a pre-planned set of drone flights and data management workflows.

This potential immediacy or near-real-time production of data brings strength in terms of fast answers, but there are risks in an open-data culture that you could accidentally share very personal information about people without their permission. This is why metadata protocols and information management plans need to be created, adopted and enforced. There are no standalone international standards (the closest being the FAIR Principles), and here lies another problem of being an early adopter: in the drone world, being a 'first ever' project is commonplace, because this technology has accelerated into business models so quickly, but the surrounding technologies and best practice is not quite 'match fit', and so we adapt to survive. This is not without its problems.

An example every Australian will be able to deeply understand is the need for fast and accurate information for bushfire management and defence. In the Australian summer of 2019–2020, bushfires ravaged huge swathes of the country. The nationwide mega-fires created ferocious storm clouds, called pyrocumulonimbus, which then caused lightning, which created further fires. It will take decades or longer for some areas to recover, and some never will, such as the ancient woodlands in Tasmania. Imagine if there had been automated systems installed that could be activated by an AI-based trigger from a thermal camera, be it on a viewing platform tower or on a larger drone, HAPS or satellite. This could have allowed the fire to be extinguished in a matter of minutes. This technology already exists in pieces, and there are likely to be global competitions and innovation exchanges around putting roboticised, immediate-response technology to use. Australia is working globally and leading the efforts around targeted fire prevention and relief; key pieces of work to watch are those involving Australia's Minderoo Foundation and XPRIZE in the USA.

Drones saving lives

Some people have ingeniously used drones to save lives. Children have been found when missing on large properties, and there are numerous case studies of hikers being found and rescued using drones. Self-rescue drones are not yet on the market, but they won't be far off. People have already tried to develop drone systems to help women walk safely home alone at night (honestly, why aren't women safe to walk home at night already?), as well as anti-house-theft drones that were trialled in Japan.

Urban firefighting drones range from those used by managers of high-rise buildings to detect who might be smoking on their balconies, and therefore potentially causing a fire hazard, through to long-hose drones that can fly and spray water at the same time, drones that can spray targeted fire retardant, and even drones that have been designed to rescue people from high-rise buildings. After the tragic Grenfell Tower fire in London in 2017, there have been calls for better use of drones and robotics to aid in rescuing people from burning buildings. At the time, the ladders that the fire service tried to use weren't long enough, and the London Fire Brigade didn't even own a drone. That frustrates me because we know how much eligible technology is out there – so, why isn't it deployed already? Why is there a wall of inertia around innovations that could save lives, particularly when we know that some of these technologies are already widely tested and trialled elsewhere in the world, either commercially or in the defence sector?

In Australia, our health and safety legislation is written in a way to oblige people to use new (and tried and tested) technology where it is fit for purpose and proven. It seems that more global knowledge sharing is needed, maybe? Conversations are happening thanks to global events and conferences that take emerging technologies and make them business as usual, such as through one of my companies: World of Drones and Robotics (Global).

During the pandemic, there were lots of ways in which drones proved useful; there were also many ways in which they could have been deployed, but there was no precedent and, therefore, no insurance to do so. Some trials were more successful than others, with companies such as Google's Wing running world-first and successful commercial delivery drone operations from Logan in Queensland.

In Canada, drones have already been used to transport human organs between hospitals in major cities, and on the other side of the country community nurses have not had to leave their posts on islands and take half a day and a couple of ferry trips to deliver COVID testing kits for analysis. It's great to see some companies taking the lead on these ethically driven business models in the healthcare sector. The remaining questions of scalability remain, though – it's easier to be the sole company on the market, with no-one competing or buzzing your drones in contested airspace.

Fun with drones

Aside from serious matters such as saving lives in a global pandemic, there are also lighthearted ways that drones are engaging with communities. If you have ever taken a 'dronie' (a 'drone selfie'), you'll know how much fun can be had with drones. You can get decent GPS-stabilised drones for a reasonable price, and it's a top gift to buy that person who has everything.

Once upon a time (circa 2018), competitive and professional drone racing was seen as the next big thing – potentially investible and the new, scalable and smarter F1. International leagues were formed that competed with each other, and millions of dollars were spent on sponsorships, promotions and branding. However, promoting gender diversity among drone racers was not seen as a priority – it was totally male-dominated – and now the whole idea of a formalised global circuit of drone racing has gone rather quiet (dead silent, in fact). The social acceptance of racing drones was also damaged when a horrible accident occurred in the UK and a poor little boy lost his eye after a four-kilogram racing drone belonging to his dad's friend crashed into his face. Whether drone racing makes a comeback after the pandemic remains to be seen, but I am not sure people are rushing back out to it.

At large events – especially events such as the Olympics and Paralympics, the Platinum Jubilee of Elizabeth II, and New Year's Eve celebrations – drones with LED lights are becoming the new fireworks. They are seen to be more environmentally sustainable and slightly more creative than traditional fireworks, but they do lack some of the whizz-bang magic. There's a bit of bombastic global rivalry happening between China and the USA when it comes to the largest drone displays: a recent record was 3281 drones flying simultaneously in Shanghai in March 2021 to celebrate the launch of a luxury car brand into the Chinese market.

The Australian film industry has benefited from our world-leading and very conducive drone regulation landscape, because allowing filmmakers to fly drones on set has enabled us to compete directly with Hollywood to attract some of the big blockbusters. You can't watch a show or movie now without seeing some form of drone footage in it, and when done well it is really impressive, especially when it involves sweeping landscapes and wilderness. A recent Great Barrier Reef documentary with Sir David Attenborough used cutting-edge Australian drone company XM2, which has since landed work on *Star Wars* films and other Hollywood projects. This demonstrates that if you can get to the front of the wave of new technologies, you can help curate the industry and generate your own niche (and then exploit it).

On the flipside of all this positive drone love is one of the most disturbing stories I have ever heard about drone technology, and it starts with a chap called Bart Jansen. Bart loved his pet cat, Orville, so much that he wanted to keep him around and have him taxidermied after his death. But he didn't stop there. Bart was a drone enthusiast, and he wanted Orville to have some fun after his death, including flying, and so he turned him into a drone. A dead cat drone. And now he's working on a dead cow drone. I am not sure I can say much more than that.

The anti-drones

Every superhero has a nemesis – what would they be without one? The anti-drone drone is a drone designed to take out other drones that are being used in nefarious or dangerous ways. In terms of technical capabilities and venture capital, the anti-drone ecosystem is growing faster than new drones are being created and marketed. It seems risk is better funded than reward.

During the 2014 G20 Brisbane summit, people were banned from carrying a drone on their person via counterterrorism legislation specifically created for the event. Drones were also banned during the Gold Coast 2018 Commonwealth Games, with police deploying anti-drone technology that looked like massive electromagnetic guns!

Some police forces in Europe have trained eagles to take out drones as a counterterrorism measure. The idea certainly has merit, but there are animal rights issues, because the birds can be injured by the spinning blades of the typically off-the-shelf, commercial multi-rotor drone systems. Other police forces use such anti-drone measures as other drones carrying massive nets and traps, drones that knock others out of the sky, and even anti-drone radar cannons – though technically it is illegal to interfere with an aircraft, so the anti-drone guns won't be coming in handbag size any time soon, to my disappointment.

And, of course, the potentially fake news headlines

We can't have a drone chapter without mentioning an incident just before Christmas in 2018 when the UK's second-busiest airport, Gatwick, was shut for days because of a supposed drone incursion into the airport airspace, causing havoc for thousands of people. The alleged drone operators were never found, and no drones were captured. There are hundreds of ways we could talk about drones for good, yet there are always negative media stories about flying platform technologies that get much better coverage and many more clicks.

Get back in the driver's seat

The laws around drones are constantly changing, so it's important you know the rules where you live and when you travel to other countries. Many well-intentioned people have found themselves in trouble after being caught operating drones illegally. Flying drones were banned in Tokyo after a protester flew a drone with radioactive sand onto the presidential palace as a protest around the nuclear accident at Fukushima. Journalists have come unstuck in places such as Myanmar and Iran for operating drones. I remember driving over the Kariba Dam between Zambia and Zimbabwe in the late 1990s, and if you even so much as flashed a camera, let alone took a picture of the dam, the security forces would be on you in a heartbeat. 'No Photography Allowed' has evolved into 'No Drone Zones', and you may see these signs in national parks or places of significant cultural significance, as well as defence sites or places of scientific interest. It is also not polite, and illegal in some places, to live-stream pictures of other people and their children to the internet without permission. If in doubt, don't fly, and always fly safely and follow the rules.

There is already research and citizen science happening in sharing drone imagery – the best one I know is the Australian Scalable Drone Cloud platform, which allows users to deliver drone imagery right into the hands of researchers via the national research infrastructure. It is funded by the Australian federal government and operated by Monash University with a number of collaborators, including the Commonwealth Science and Industrial Research Organisation (CSIRO) and Australian National University.

7

Frankenfood and robotic rabbits

The future of what we eat

'You may go into the fields or down the lane, but don't go into Mr McGregor's garden: your father had an accident there; he was put in a pie by Mrs McGregor.'
– Beatrix Potter, *The Tale of Peter Rabbit*

There's a key concept when it comes to ecology and the survival of species called 'carrying capacity'. This is the notion that each environment has a set limit of how many individuals of a species it can accommodate with the resources available. Humans broke this rule the day that agriculture was invented: as soon as we were able to manipulate the amount of food and other resources that could be produced, we changed the rules and stretched the natural limit. On a planetary scale, we're at risk of being overcome by factors such as disease, genetic modification and climate change. And still we continue to pursue two things to an extreme: one is change and innovation, and the other is keeping things exactly as they have always been.

Agtech (agricultural technology) is a growing sector in many developed economies and a huge area of investment and opportunity, particularly in the Antipodes. Over the past 10 years it's been valued globally at US$189 billion. The sector is ripe for technological disruption because it is on the bleeding edge of a number of existing and emergent crises, including weather disruptions, biosecurity issues, supply chain evolution, reduction in the number of seasonal workers, and all of the worst of the climate crisis – more extreme weather events, floods, droughts, fires and human pressures such as the demand for produce that looks 'perfect'.

So, how will we feed the extra billion mouths that are due to arrive in the next few years? How can we farm more, waste less, spend less, and create fewer climate-changing emissions while doing it? The answer is definitely not 'how we have always done it'. It lies in developing and funding multimodal, cutting-edge technologies to work on the land, in the sea, and in anthropogenic, closed-loop urban farming solutions.

Feast or famine

Experts estimate that in the Western world we throw out about a third of all the food we produce. This is nigh-on criminal. This shocking statistic has led to public outcry about the standards imposed by the supermarket chains, expecting our farmers to only produce fruits and vegetables that are of uniform size and shape. At one end of the scale, we have a billion people on the planet unable to get enough food; at the other, we have a billion overweight people who are overloaded with calories but possibly still malnourished in terms of essential nutrients. We are also still living in a world in which famine continues to be used as a weapon of war.

One of my earliest childhood memories is of the awful famine in Ethiopia in the 1980s and the subsequent Live Aid concerts and Band

Aid records to raise money to get food into the mouths of starving people. The BBC news bulletin by Michael Buerk started with these words:

'Dawn, and as the sun breaks through the piercing chill of night on the plain outside Korem, it lights up a biblical famine, now, in the 20th century. This place, say workers here, is the closest thing to hell on Earth. Thousands of wasted people are coming here for help. Many find only death.'

It turned out that this famine wasn't solely about low annual rainfall and the subsequent failed harvests at all, but also war and politics. I can still remember the video and photo coverage of that time, and the general sense that people in the West wanted to help. But food and geopolitics, it seems, are inextricably linked. There has to be another way to ensure that no-one else starves to death because of conflict. Unfortunately, though, as I write this, famine is raging in Yemen and being exacerbated by conflict.

So, how can technology help prevent famine? Well, the way in which we grow food is going to substantially change in the next few years, and one of the largest hurdles to advancing food security globally is the adaptation and adoption of new technologies in different countries. Imagine the local knowledge that should be applied when adapting new methodology, and also the new power and water systems that will be aligned to new methods of 'smart farming'. This isn't calling existing methodologies 'stupid farming', but the reality is that new methodologies are available that were not available to farmers over the past few thousand years, including the advent of precision agriculture, data science and AI-backed operations.

One of the most pressing issues in current agricultural practice is sustainable soil management. Soil is a magical thing deserving of its very own book. My PhD was on soil microbial ecology – the relationship between bacteria, fungi, actinomycetes and other tiny organisms that function in the soil nutrient flow and the food webs that keep everything running. I looked at how to model the function of soils in terms of change management, with a view to creating predictive statistical models of behaviour. Some recent studies have described current soil management methodologies as irresponsible and linear in their thinking, with one study calculating that under current techniques we have only about 60 harvests left on a global scale. This would mean that, if we keep going the way we are going, there will be global famine within my children's lifetime. We should be worried.

The oceans are also under pressure, suffering issues of overfishing, dead zones (areas that have too little oxygen to support marine life) and plastic, but they will be expected to provide more food. More novel methods are being touted of sustainable aquaculture of protein sources including oysters, abalone, mussels and other fast-growing seafood. There's also a growing market for macroalgae such as kelp, with kelp burgers now becoming very fashionable in California.

Legislation, politics and 'mad cow' moments

We need to talk more about life cycle analysis (LCA) when it comes to the food we put on our plates. Some purported food solutions are actually worse for us and the environment once the entire life cycle of creating, transporting and disposing of the product's waste is considered. Some 20 years ago, the big push was for low food miles, encouraging the wealthy and those with choices to look locally for produce. But in the UK, buying locally reared lamb actually had a higher carbon footprint than eating New Zealand lamb (slaughtered and deep frozen before distribution, so none of the ethics of live transport applied) despite the shipping emissions, due to the manner in which the UK lambs were reared and the extra feed they were provided, which has to be grown. So, if a reduction in carbon footprint was the main goal, then New Zealand lamb would have been the answer. However, if the desire was to support local food security and the agriculture sector, then local lamb would have been it. There is no clear right or wrong answer to the choices within our complex and global food supply chains – we just have to do the best we can with the information provided at the time.

If we want our food to reflect our environmental codes and moral ethics, we need to be prepared to pay for it – we need to vote with our shopping trolley before it is too late for those in industry who are trying to the do the right thing (or the lesser of two evils). An example of this is the UK's pork industry, which has become more expensive (and therefore less competitive), partially because of animal rights legislation that was put in place in the UK that was not applied across the European Union (EU) – for example, the cruel practice of sow stalls is banned in the UK. There has been a recent call in the UK to boycott Parma ham because of welfare standards.

There are also other politically charged issues in agriculture, such as the EU's decades-old common agricultural policy (CAP), which pays subsidies to farmers to keep them farming their land regardless of sustainability issues. This policy-driven culture is believed to have been a fundamental factor in the 2001 outbreak of foot-and-mouth disease (FMD) in the UK, because one way the disease spread was at markets where people were borrowing and mixing up each other's animals for head counts to apply for grants.

If you're around my age or older, you'll remember the outbreak of bovine spongiform encephalopathy (BSE) – 'mad cow disease' – in the 1990s. Just as Spain was associated with the Spanish flu, the UK was associated with the prion protein disease in humans variant Creutzfeldt-Jakob disease (vCJD), when it is now thought that it was a spontaneous variant that was found all over Europe at that time, with a significant cluster also arising in Spain. BSE is still frequently detected in cattle around the world (and there are 1.5 billion cows on this planet), but it rarely makes the news because the human disease is not as prevalent. But watch this space – I bet we will see headlines about prion protein diseases again this decade.

Eating marine-grown seafood and fish, and therefore ingesting the bioaccumulated plastic they have eaten, may be the media's next 'mad cow' moment, because nanocarbon is known to be able to enter our cells and move around our bodies in ways we are just beginning to understand. What health crisis will come about from eating plastics? Only time (and science) will tell, but it's more likely than not to be bad news, unfortunately.

Pink slime, mock meat and creepy-crawlies

Would you drink synthetic milk? What if you couldn't taste the difference between cows' milk from a farm and synthetic cows' milk from a lab? A lot of headlines have been generated by the advent of lab-based chicken nuggets, but not so many about some of the other consumables that are currently being created, tested and approved – synthetic versions of natural fibres such as silk (effectively silk without the worms) and 3D-printed, lab-grown meat established from stem cell lines (steak without a cow).

During the 1990s, there was a huge pushback against genetically modified (GM) plants advancing to field trials. They were even called 'Frankenfoods' – a play on Frankenstein's monster, who was patched together from different dead people. Nowadays, we use GM crops without really hearing or knowing about it, but the idea of messing with the genetics of plants was enough to cause public outcry at the time. How will people perceive lab-grown, processed, printed, stem-cell-based lines of meat, and how will it be marketed by the modern (and social) media? As someone who doesn't eat red meat, I am in a conundrum about what to do with a 3D-printed steak; I think the scientist in me would want to give it a try, and I would be extra happy that no animals were harmed in the making of my steak.

Cultured meat – also known as 'cultivated meat', 'clean meat' or 'cell-based meat' – is produced using tissue engineering techniques traditionally used in regenerative medicines. In 2013, Professor Mark Post at Maastricht University pioneered a proof of concept for cultured meat by creating the first hamburger patty grown directly from cells. Since then, other cultured meat prototypes have gained media attention: SuperMeat opened a laboratory restaurant called 'The Chicken' in Tel Aviv, Israel, to test consumer reaction to its 'chicken' burger, while the 'world's first commercial sale of cell-cultured meat' occurred in December 2020 at Singapore restaurant 1880.

People buy processed meat products made up of the 'pink slime' parts of traditional meat processing (where all the smallest offcuts of the meat are blitzed together and liquefied before being reformed as chicken twizzlers or similar). If you can eat fast food products like these (and feed them to your children), which use the lowest-quality remnants of the meat production process, then surely you wouldn't mind your 'pink slime' coming from a bioreactor in a lovely, clean lab with all the rules and regulations that control it?

Mock meat – meat-like products made from plant-based ingredients – is another massive area of growth in the food industry. There are environmental and health concerns around these highly processed products – especially those that contain soy or palm oil, both of which are directly responsible for forest destruction around the world. Gluten-based seitan mock meats can be high in salt and fats (in an effort to find that magic taste balance), which are not necessarily good for your health. Soy goes into vegetarian burgers, of course, but the majority of soy grown in places such as Brazil is there because it is part of the animal-feed supply chain. So, we are growing soy and maize not for human consumption but rather to export as animal feed to other parts of the world.

After having been a slow burner for a few decades, veganism is finally upon us as a solid option. Believe me, trying to be a vegan in the 1990s was pretty hard (thinking about all the Sosmix and bean burgers I suffered makes me shudder). Now veganism has gained influencer status, with music powerhouses such as Beyoncé and Adele publicly talking about their vegan diets. Those for whom veganism is a bit extreme are finding ways to be flexitarian, where they diversify what they'd normally eat and choose more plant-based options. Many people who love meat and will always continue to eat meat are also reducing their meat consumption – becoming reducetarians, who actively reduce the animal products they wear and consume, including leather and honey. In 2009, Paul McCartney and his daughters Mary and Stella started a movement called Meat Free Monday, which has taken conversations about what we are eating to dinner tables around the world. (I do have a thing for the sausage rolls from the Linda McCartney vegetarian food range with tomato ketchup. Yum!)

CEOs have given TED Talks espousing the value of a meat-free week, and restaurants have instituted meat-free menus. There's even an annual 'Veganuary' (vegan January). It will be exciting to see what the increasingly scientific celebrity chefs of the world do with their Michelin-starred creative brains to help produce the next food craze. My bet is on something to do with kelp. Who knows what else we could farm in marine systems that don't carry the environmental burden of large-scale aquaculture (such as salmon farming does)?

Whether we all embrace veganism or not, we know that 'business as usual' will no longer work in the future of protein production. It is ethically and environmentally impossible to feed the whole of the world the levels of protein that those of us in the West have grown accustomed to. The answer may be in insect-based protein – the thought of which may make you shudder, but stay with me on this.

Have you ever been to a South-East Asian country where people eat insects as a delicacy? I remember being culturally shocked as a young British woman doing my Year in Industry in Thailand and seeing people go to the bug stand for snacks. I'm actually disappointed in myself that I never tried them at the time – I'd lost my gumption!

Insects are nutrient-efficient compared to other meat sources. Crickets, for example, contain a similar amount of protein as soybeans and include mostly unsaturated fat. They also contain dietary fibre and some vitamins, such as vitamin B12, riboflavin and vitamin A, as well as essential minerals. Locusts contain 8 to 20 milligrams of iron for every 100 grams of raw locust, whereas beef contains roughly 6 milligrams of iron per 100 grams. So, you can see the macronutrient and micronutrient status of insects balances out against what we'd expect from meat or meat substitutes.

There are seven insect species that have been targeted as suitable for mass production and human consumption: mealworms, lesser mealworms, house crickets, tropical house crickets, European migratory locusts, black soldier flies and houseflies. It's unlikely the average Westerner will chow down on crispy bugs for lunch, so I predict that instead a much more acceptable form of finely milled insect protein will be hidden in the form of protein shakes, high-protein flour, or as a base for soups or stews. I would definitely use insect protein powder much more readily than chewing on an exoskeleton. Would you?

Farms of the future

As human population centres become more and more urbanised and we are faced with an acceleration in the growth of megacities – in part due to the effects of climate change – new problems emerge. How do we grow more food when we have less land, and when the oceans and natural waterways are far away? One of the solutions is urban agriculture, which sounds oxymoronic, but it is a growing sector already. The Netherlands has been a huge leader in this space. Their poor-quality, reclaimed-land polder soils are unproductive (and have a distinct lack of earthworms), and there isn't the space to feed a growing population using traditional methods. With sea levels rising, some of their land may not be accessible at all in the next decade. These converging challenges have motivated the Dutch to invest heavily in closed-loop farming, where pigs, fish and vegetables are grown in an industrially connected mega-warehouse system. The aquatic animals and micro-ecosystem clean the wastewater, and the pigs produce the fertiliser and get fed at the same time.

Some well-funded start-ups are working within the global tech world to utilise large warehouse facilities to create vertical farms without pigs, in which vegetables and other plants are grown on hydroponic tables with many layers. They are automated in terms of their fertilisers and water use, and some of them don't even use soil anymore. These AI-driven precision farming techniques can be used en masse in a city location, possibly even downtown. We can farm on brownfield sites, such as former ports or storage yards, old malls, or areas of demolished low-cost housing. In Australia, vertical farms could also be established in regional areas to prevent the inflated pricing that happens when fresh fruit and vegetables are transported to remote communities. Can you think of any spaces around where you live that could be better used as spaces for vertical farming?

Urban farms are automating more and more, and diversifying the crops they produce – even using robotic drone 'bees' to pollinate plants. Robotic drone 'bees' might sound ridiculous, but we are facing a global bee crisis, and microrobotics are a viable technical alternative. Until June 2022, Australia was the only major honey-producing country free from varroa mite, but now there is a biosecurity battle underway to eradicate an incursion in New South Wales, and I hope they eradicate it sooner rather than later. The death of the beautiful bee would mean massive food shortages and certainly less variability in the food we enjoy today. Imagine: no more stone fruit, apples or honey. In countries where bees have already disappeared, humans can get jobs as pollinators. With little bags of pollen and a paintbrush, they climb ladders and walk miles to pollinate plants that a bee would normally pollinate. This is expensive, of course, if you are in a country with ethical labour laws, minimum wages and workers' rights. If tiny drones can fly around like bees and pollinate, well, labour laws and daylight restrictions do not apply to robots.

Soon we'll be eating fresh fruit and vegetables that have never been touched by a human hand from plant to plate. The rise of the roboticised and automated farm is upon us. AI-backed technologies have already proven to be better than humans at aspects such as grading and packing avocados, picking and packing raspberries, and sorting shopping. Amazon holds global robotics competitions at which university researchers create robotic and automated systems that then compete as identification, picking and packing robots, and most are already more accurate and efficient than the humans who created them. Add automation to the types of plants that are grown, and soon we will have an agtech convergence, which will mean we will never again have labour shortages of fruit pickers causing price hikes, or food rotting in the fields because people can't get to it. This is what robots are best at: doing robotic, relentless, repetitive, boring tasks, and especially ones that may come with health and safety issues.

We will see farmers wearing exoskeletons, robotic sheep dogs, and autonomous drones checking fences and counting heads of livestock. We already have satellite tags for cattle to prevent livestock theft (yes, that actually happens) and dairies that are fully automated to milk the cows when they choose to be milked. Nary a human will need to be nearby – but humans will still be involved, likely in remote operations and information management, as well as on-site maintenance. The suggestion that robots, drones and AI will remove all the people working in agriculture is a fallacy put forward by those who are scared to see modernisation and technologically driven solutions maximising food production (remind you of Ned Ludd?).

The rise of the smart farm has also meant a rise in the amount of data being collected, collated, analysed and stored. There is scope for microanalysis on farms, in which farmers can analyse metre by metre the productivity of a particular piece of land, and robots can look centimetre by centimetre for weeds to extract. There are also large-scale satellites that can look at general capabilities of the land to be used for agricultural production, such as soil carbon, erosion, water use and supply. Anyone who works in soil science knows this means huge generalisations, and the law of averages needs to be applied to natural systems when modelling potential land use, but these data sets are used more and more to predict agricultural viability and capacity.

Some really interesting studies and spin-off businesses in Australia are going global in this area of smart farms. One of these is The Yield, which has recently entered the US market. It uses artificial intelligence to take local weather and environmental data, as well as field produc-tivity down to a centimetre-level scale, in order to provide information about what should be planted and how it should be managed to maxi-mise yields. Aside from the AI systems, some agricultural drones now have onboard processing, in which the drone itself analyses the field or paddock as it flies, and then does a return flight to spray particular fertilisers as needed, or maybe even precision-spray herbicides.

Protecting provenance

Once we produce all these wonderful agricultural products, we need to prove they are what we say they are when we export them into global markets. As mentioned previously, there's been a lot of work done, especially here in Australia, looking at ways in which food provenance can be proven using innovations such as radio-frequency identification (RFID), hiding really small microdot identifiers, and even DNA testing. There are also regional airports developed purely for the export market – such as in Toowoomba, west of Brisbane – and these are proving to be incredibly lucrative for those involved, adding value to the provenance of goods and produce.

The market for high-quality produce, particularly that which is ranked by the location it is produced in, means we have to find new technical and legal ways to fight the battle against food fakery. But how do you protect food provenance through legal mechanisms such as patents or trademarks? The French have been doing this for years with products such as champagne (it has to come from the Champagne region in France to be called 'champagne'; otherwise, it is a 'sparkling wine'). If we can be assured of the provenance of the products that we are buying, we can make consumer-power choices to make sure we do not buy food products that have a significantly negative effect on the local environment or the people who are producing it. This gives more power back to the consumer but also places ethical, social, moral, economic and legal obligations on the board directors of the companies producing these foodstuffs to make sure they are not causing deforestation in the rainforests of the world, or unnecessarily polluting (such as causing fertiliser runoff into the ocean), or have slaves involved in the supply chain. ESG investing has turned to the global food industry, and a closer lens will be placed on where things come from.

Growing your own

Those of us who have the money, space and time can produce at least some of our own food. I know more and more people who are transporting part of their gardens back to the late 1940s, when people grew most of their own vegetables in their very own 'Victory gardens'. When I lived in the north-east of England 20 years ago, gardeners were competing for space on government-owned land and veggie patches, also known as 'allotments', where they could grow their own produce. The same is happening across Australia today, with community gardens located in inner-city suburbs and even regional towns for those who don't have the space to grow their own at home.

Cottage farming is the new artisan. The rise of sourdough bread baking during the pandemic is one example of how people are being reminded of our innate connection with the food we make and eat. Awareness of where food comes from is something that should be taught in schools. So should cooking, so that children grow up with the basic ability to cook things from scratch or from whatever is in the cupboard and fridge. This would allow more people to access food economically. Also, where are the high-tech, AI-empowered community gardens?

Get back in the driver's seat

Do you look at the label when you buy produce from the supermarket? Do you shop for it online? How often do you get to a farmers' market, or maybe drive past a produce stand? There are many ways we can use our purchasing power when it comes to the source, supply chain and sustainability of the foodstuffs we choose.

An easy place to start is to not let any food go to waste. In Australia, we discard more than 30 per cent of all the food that we buy, and the carbon footprint caused by the landfill we create is more than we create through the aviation industry. We need to stop wasting food and start paying the right price for produce – or, better still, get the government to provide grants to producers who do the right thing. Between planning your food for the week, avoiding plastic packaging and preventing waste through different recipes or storage methods, there is a lot we can do as individuals that can add up to a big collective difference.

You could also check out charities such as Foodbank Australia and OzHarvest Australia, and maybe buy a meal for a family you'll never meet. The cost of living is becoming a real issue, and the use of food charities is increasing.

Lots of food for thought.

8

Digital doppelgängers

The future of health

'There is no gene for the human spirit.'
– Tagline for the movie *Gattaca*

In the next few years, it is highly likely that you will meet your digital twin: a virtual version of you that lives online, on data servers and in cyberspace. Your digital twin will be made up of many different inputs – from your medical records to your social media profiles, location data off your phone and smart watch, and even surveillance camera footage. Digital twins are currently being built of buildings and cities, but the ultimate project is producing accurate digital twins of humans.

The purposes of human digital twins are diverse and broad, and really depend on the companies and entities that are building them. For example, a digital twin could include data that describes your body and behaviour, which could be used if you become unwell and need medical intervention, helping medical experts prescribe you personalised and bespoke treatments. Digital twins of people across our communities could also be used collectively to help improve our understanding of public health issues, such as pandemics.

More personalised medicine – new therapeutics and medicines, as well as new treatments for specific diseases – generally means personalised solutions and better results. The use of tailored immunotherapy treatments such as chimeric antigen receptor (CAR) T-cell therapy will likely mean that one day, chemotherapy will be a thing of the past. We are seeing speedy shifts in synthetic biology, and soon those of us who remember the start of the HIV/AIDS pandemic will rejoice to finally see the virus be sent to the archives of history. We'll also be better prepared for the next pandemic, because we will be able to get a vaccine going in days rather than years.

The internet of bodies is the next stage of the future of health. During the fourth industrial revolution, the internet of things was the idea of connecting all the sensors on all the platforms we have – cameras, drones, CCTV, smart devices, environmental monitoring stations, and so on – and then transforming that data so it could help us make sense of the world, how we live in it, and how we could predict the issues that come next. The internet of bodies will allow us to do this from a public (and personalised) health perspective, from the scale of a single cell to a whole digital doppelgänger of you.

The future of robotics does not look like cute, white plastic, social, humanoid robots serving us tea and cake, but rather technological advancement inside our own human bodies. It will happen on a diverse scale – ranging from an organ to an entire joint system – for the purpose of making our lives better. Technology will influence our physical and mental health, changing how we live, work and play.

Change is fast, yet slow

As we swim into the fifth industrial revolution, we are being greeted by the power and tyranny of multimodal data that seeks to understand who we are and what we are made of. When data is collated from different sources like this, it can be problematic. It can be augmented, filtered, personalised, depersonalised, adapted or even deleted. Most people in the medical world are not trained as data scientists, and a lot of systems still rely on pens and paper rather than cutting-edge IT systems. I am amazed every time I have to go into hospital by just how old-school the place feels: it's certainly not what the sci-fi movies said hospitals would be like in the 21st century.

When I was pregnant with my first child, I ordered a copy of my ultrasound images. Instead of a USB stick, I was given a CD. This was 2017! As an Apple user, I didn't have a disc drive, so I had to dig out a really old laptop and hope it worked so I could see the scans digitally. I ended up taking photos of the printouts they had given me instead of trying to decipher the bespoke medical software. I was amazed how behind the times they were – and many parts of the health system still are. It seems that while many areas of medicine are cutting-edge, a lot of the systems that support this work are decades older than they should be. It's unclear whether this is a funding issue or just a reflection of the slow pace of change that comes naturally to the world of healthcare: move too fast in medical trials or drug development and you'll be called a danger to society. This is a huge inhibitor to innovation and disease prevention.

So, how on earth are we going to ensure that converging technologies and data sets work smoothly, especially when it comes to human-based data? The answer lies in big tech, I'm afraid. There are competing and overlapping markets occurring around the data sets that are being used to create digital doppelgängers. Companies are racing to create the best software and hardware that can be seamlessly and ethically used, in a confidential and cyber-secure way, for you to start building the digital version of yourself by proxy as you use social media and other health tech. Remember this: if an app is free, then *you* are the product, and it doesn't get more personal than when we are talking about your body. Who would you trust more: the government or big tech?

The commercialisation problem

I felt inspired and then completely enraged when I learned about a simple test that had been developed by a local university. It could pre-screen cancer patients to see which chemotherapy treatments would work best for them with the fewest side effects. However, it wouldn't be developed for use because the university couldn't get funding for it. The total cost of that test per patient? About AU$30. That's not much more than the price of avocado on toast and a coffee at your local cafe. Why on earth was it not proceeding?

This is the problem with new and emerging health technologies: there are so many restrictions preventing them from scaling up to a profitable size. In Australia, we are known for our research and development, but our commercialisation is not exactly world-leading. We just don't have the cashed-up entrepreneurs or a large enough population to have a significant amount of capital going into spin-offs and start-ups.

Deep tech and synthetic biology are two major areas of growth. Deep tech (also known as 'hard tech') companies have the expressed objective of providing technological solutions to substantial scientific or engineering challenges. These challenges require lengthy research and development phases, and large capital investment. The most prominent deep tech fields include advanced materials, advanced manufacturing, artificial intelligence, biotechnology, blockchain, robotics, photonics, electronics and quantum computing.

If you had an mRNA-based COVID-19 vaccine, you have benefited from synthetic biology. It's when scientists create molecules in the lab that are produced inside your body's cells in order to trigger a function inside those cells. It's amazing (and not cheap) work.

Convergence is a significant part of these areas of study, with many multimodal data pools being combined and aligned. Herein lies a real challenge and opportunity: if we really want truly personalised medicine, then we need to invest in the data and information systems required to make it fit for purpose, rather than trying to retrofit data sets with no metadata or ISO standards applied to the capture. Realising the concept of collating information from sources as diverse as your Apple Watch and a bespoke medical device in your heart, for example, will take a lot of development.

It takes an ecosystem to nurture the new and emerging seedlings of the tech world – especially when it comes to highly risky research and development phases, where a lot of time and energy is invested long before significant dollars can be recouped. Being problem-focused rather than sales-driven means there may be a longer lag phase to any return on investment, and markets may shift as technologies are developed. Pandemics and other such incidents can cause a rush of money and a speeding up of the cycles needed to safely get results, though honestly it would be better if we didn't need a pandemic as a motivator to trigger government spending. The way science is funded needs a shake-up, and maybe a target of a percentage of GDP – let's start with 5 per cent here in Australia, please.

The machine inside you

Would you wear a tracking device or chip under your skin? Some people love this idea, and people have even been known to take the chips out of travel or credit cards and surgically insert them into their hands or wrists, so they don't need to carry anything with them. This may sound fringe and fantastical, but it's on the increase. In the 2012 remake of *Total Recall*, the lead character has a phone implanted under the skin of their hand, and they literally put their hand to their ear to make a call (much like my two-year-old does during games of make-believe). How handy would it be if you couldn't lose your phone, or drop it down the loo, or have a toddler hide it so well it can't be found for days on end? As a woman, having a tracking device under my skin might also make me feel safer, knowing that I would be trackable if something terrible happened. But what about women fleeing domestic violence who don't want their abusive exes to be able to track them down?

Technology is bringing much more than just convenience to the medical field: dreams are now becoming realities. Kids born with microtia (where their external ear is improperly formed) can have a new ear grown in a lab using their own tissues through innovative additive manufacturing techniques (aka 3D printing), which can then be surgically attached. Now, imagine the possibilities if you could integrate hearing assistance into the ear you've just grown for them, instead of just aesthetics. Why aim to merely replace when you could improve and augment?

Researchers are working on creating stem cells to combat various diseases and disabilities, and for tissue replacement or repair. One of the most exciting areas is spinal cord injury. Stem cells taken from the nose can be turned into nerve cells and then injected into the spine. Add this to immersive VR headsets and treadmills, and you create a whole new multimodal rehabilitation opportunity. Effectively, when spinal patients look down in the VR world, they will be able to see their legs walking – so their brain thinks they're walking. We know about neuroplasticity – the ability of the brain and nervous system to change. It is an area that we're really just getting to grips with, but in these spinal cord patients these innovations have created a whole new level of physical therapy: people who may have been paralysed for life are going to be able to walk again.

This is another example of how convergent technologies add more value than the sum of their parts. It makes me wonder what would happen if you took an exoskeleton, a treadmill, a physical therapist, a neuroscientist, a VR expert, a computer scientist, an artist and a gamer, and put them all in a room with a spinal injury patient and teach them what outcomes might be possible and preferable for the patient.

We are reaching the fulcrum of additive manufacturing of human tissue. Life-changing support is already available in the form of bone grafts made to order and bone replacements designed from MRI scans to be completely fit for use and personalised – like how you can have clothes tailor-made for your own dimensions. If you could have a back-up body made up of parts available for use when your current body breaks down, would you? This is a key concept of the 'hospital of the future', and I first heard about it from Professor Mia Woodruff in Queensland, whose work in this field has begun being scaled and applied to patients in Australia.

If we can grow cells, tissues and maybe even organs to order, what happens if we look subcellular? Let's go nano! Nanotechnology (one nanometre is one billionth of a metre, so this is technology operating at the scale of molecules) has been a headline-grabber since the late 1990s, but was first introduced as a concept in 1959 by physicist Richard Feynman in his world-famous lecture 'Plenty of room at the bottom'. It's estimated that as a sub-industry, nanomedicine (the medical application of nanotechnology) will be worth US$260 billion by 2025. We have come a long way from people building replicas of famous statues at the nanoscale to grab media attention.

Mother Nature has the answers

As always, Mother Nature has already engineered answers; it's just that we have yet to realise the potential. There are places in the ocean that are so unfamiliar to Homo sapiens that we haven't adapted to them, and our immune systems haven't evolved alongside them. They are, by very definition, alien worlds to us humans and other land dwellers. Some of the microscopic life that resides there could be used as part of the next drug delivery mechanism, and developed into new medicines and therapeutics. These are called 'silent bacteria', because our immune system and other cellular defence mechanisms just can't detect them. They might currently seem like the stuff of sci-fi movies – maybe slightly reminiscent of the 1971 movie *The Andromeda Strain* to some of you – but the more we seek them, the more we find that Mother Nature has already come up with the solutions we will need to survive on this planet for the next few thousand generations.

Imagine if, rather than just waiting for a solution, we could use big data alongside bioinformatics and actually switch on silent genes to make regular bacteria produce new proteins (or medicines, or antibiotics) in a bioreactor lab? There is work happening now to look for new antibiotics; indeed, we are racing to find them, with the search going as far as to the depths of the Mariana Trench and the heights of the International Space Station. The next pandemic is likely to be driven by antibiotic resistance, and so we need different weapons in our armoury. Antibiotic use in humans is only 100 years old, and we have squandered the potential of these miraculous molecules by overusing them, not using them properly and, in some countries, not protecting the last line of defence by using antibiotics reserved for humans in pig farming and other forms of intensive agriculture. What else could we find that is locked up in these silent biosynthetic gene clusters (BSGs) – parts of the microbial genomes that are sitting there quietly, not working or producing anything at the moment – when we start to stimulate them? It's a potential treasure trove.

When we can better understand Mother Nature's ways on a molecular level, we will surely be able to imitate in the lab how our bodies function – right? Scientists are already working on new ways to understand cellular processes and then use them to produce proteins, much like hijacking a factory that normally makes shoes to instead make leather handbags. Synthetic biology is an industry worth some

US$25 billion and is likely to be the great scientific gift from my generation – and what a cracker it is! mRNA is the code that asks our cells to make certain proteins, and the vaccines for COVID-19 used the mRNA pattern to make the spike proteins that the virus uses to unlock our cells and cause infection. By creating the spike proteins inside our bodies for a very brief period of time, we were able to train our immune systems to recognise what they would look like and so have a defence mechanism ready to go when faced with the real thing.

Following the success of Hungarian biochemist Katalin Karikó (who deserves a Nobel Prize), we are now recognising not only the potential of new technologies but also the crazy pitfalls and politics around academic funding of emerging research. Katalin's work will save millions of lives in the next few years alone, but the most exciting impacts of mRNA-based vaccine technology are yet to be truly discovered. There are now mRNA vaccines being created, tested and trialled for diseases such as dengue fever, malaria, zika, and even HIV – diseases which typically affect poor people in the Global South and so have not traditionally had a lot of money spent on them.

The transcriptome is a fascinating beast. Essentially, it's a descriptor of ribonucleic acid (RNA) activity. RNA is a molecule that comes in many forms, with various biological roles in coding, decoding, regulation and expression of genes. To understand RNA, let's break down the main difference between DNA and RNA. Imagine your cells are libraries, and the books are stories written in DNA. They silently sit there waiting for someone to open the book and read them. When the DNA is read aloud, that sound is the mRNA (messenger RNA), which disappears almost as quickly as it was formed, because it has a simple and short-lived reason for existence: to carry a message which confers a function, such as protein expression.

Since the start of COVID-19, we have heard more about mRNA than I have had hot dinners, but the interesting thing about the transcriptome is that it is at the convergence of molecular biology, microscopy, big data and creating digital twins of ourselves. An area on the horizon of research is single-cell spatial transcriptomics – that is, the ability to look at that book being read in real time and see the RNA being produced on a single-cell level, so that we can then understand cell function at an organelle level. What this could mean, when the area of study reaches its potential, is that we would be able to personalise medicine, surgery and drug doses. Imagine if, before a surgeon operated on someone, they could operate on the patient's particular problem in a simulated environment and see the outcomes from different approaches.

I remember at the start of the pandemic all the mainstream media outlets rushing to state that we had never managed to have a vaccine against a coronavirus, but our successes in this regard irrefutably show that it is only money that is stopping the science. Now we have a new dawn in terms of the value of public health, and hopefully that will be matched by long-term funding strategies and the reconnection of people who have fallen away from their STEM careers to come back to a place where they can be celebrated and supported. Thank goodness for the gumption of Katalin Karikó, Sarah Gilbert and the other leaders developing new vaccine technologies. They are the heroes of the early 21st century.

The human genome

You may remember the hype around the human genome project when it started way back in 1990 with an aim to map all the genes of the human genome. It had 20 major collaborating partners from all over the world. An initial draft of the human genome was shared in 2000 and the project was stated to be complete in 2003. What used to take years now can take minutes, and the growth of genomics to be applied to other species, as well as whole ecosystems, has really come into its own in recent years. The technology just keeps getting cheaper, and business models are diversifying.

To many nowadays it may seem like the world has always known about genetics, but the accessibility and ease of testing is still wondrous to me. I recently had my genetics tested because of a family history of cancer, and it was a surprisingly easy process. All I did was spit into a tube and post it to California. For a few hundred dollars, I was very happy to receive the news after a week or so that I had no gene variants that could mean a predisposition to cancer. This wasn't just important to me but to my sons: some of the genes associated with breast cancer (such as BRCA2) are also associated with prostate cancer, so I wanted to check that my boys would be OK in terms of genetically related cancers.

Your genome is not actually a static, set-in-stone thing; it is dynamic, and the study of epigenetics (the interaction between your genes and the environment) has really gained traction, investment and media coverage this century. You may have a genetic predisposition to certain illnesses, for example, but that doesn't mean you are definitely going to get a specific disease. However, what it does mean is that certain stimuli and environmental factors might activate some of these genes and then cause a problem for you. Many such genes might be inherited, so you might have a family history of a certain type of disease (such as cancer or Alzheimer's), but not all cancers or neurodegenerative diseases are directly linked to inherited genetics. It's all part of a puzzle we still haven't cracked: nature versus nurture.

The next area of science commercialisation is of the proteome – the analysis of the proteins that your cells are currently producing, what they could produce, and which functions they perform. Then add in the metabolome (the collection of all metabolic chemical compounds produced by cells during metabolism) and you're looking at an ecosystem of unique and highly complex small-molecule components that you make and that affect your life. The health and fitness industry is already talking about personalised and targeted nutrition and exercise plans based on genetics, but these tailored solutions will really come into their own when we can crack and master the metabolome cheaply on a person-by-person basis.

The more we understand our genetic potential, the more we can understand the biochemistry of disease potential in our own unique biometric footprint. There are huge demands for better testing to detect (and predict the risk of) disease formation. In the UK, cancer screening using blood samples has started on a population of 140,000 volunteers and will ramp up to a million in a couple of years. In Sydney, Dr Caroline Ford and her team are creating a blood test to cheaply and reliably detect ovarian cancer in its earliest stages. She is using her deep knowledge of cancer cell biology to weaponise its behaviour into a mechanism to test for its presence. Genius, really.

The microbiome

An area of study emerging through the microbial ecology community in the early 2000s was that of the human microbiome, which consists of all the microbes in and on the human body. We have more bacterial cells in our body than our own cells. You and I are truly living in a microbial world, and I am a microbial girl.

The very idea that our gut microbiology could be attached to so many chronic health issues, as well as mental health and performance, and potentially be the cure of some persistent diseases is very new, but it is gaining traction in the media and in general medical practice. And, no, it is not merely fecal matter. Our gut health is fundamental to our mental and physical wellbeing. You can even have your gut microbes commercially analysed now and receive tailored advice on how out of balance you are and what to do to fix it – it will cost you a few hundred dollars, though, through a University of Queensland spin-off called Microba.

Catastrophic collapse of a person's gut microbe community can lead to some really negative health outcomes. Sometimes this is triggered by really strong antibiotics; other times it could be from hospital-acquired infections and superbugs. One answer to this is a fecal microbiota transplant (FMT), where someone donates their poo to be pumped into your gut to help re-establish gut flora. You are unique, and there are gut microbe species in your body that can only be found in your body and no-one else's – so the key seems to be to use the poo of someone who is closely related to you. (Imagine that conversation!) I have had friends whose lives were changed for the better by FMT, so they talk about it all the time – it's great geek chitchat, but not so great at a dinner party.

Get back in the driver's seat

The future is an incredibly personalised space, which means it is very, *very* personal in all the ways we can find awkward. With some of the advances in medicine focusing on tailored solutions, it could be said that we don't have the science literacy in society for the general population to fully comprehend how some of these technologies work, and thereby give true consent to what they are about to go through.

The key here is not only for us to take responsibility for what we know and can learn, but also a societal responsibility to trust science and ensure that new technology is regulated appropriately. Get educated, and ask your local MP to do the same; ultimately, rules and regulations are made in the political sphere.

Don't spend too long on 'Dr Google'. Ask better questions. All medical doctors take the Hippocratic Oath or its modern successor, the Declaration of Geneva, which means they make promises such as to do no harm and always make the health and wellbeing of their patient their first consideration. But people are human, and humans can make mistakes. Always ask for a second opinion from someone unconnected to your original doctor.

If you have the money, get your microbiome checked. You'll likely enjoy the results, even if only for dinner-party conversation fodder. Also, if you have a history of cancer in your family, get a referral for genetic counselling and look at the options open to you. I recently had my first colonoscopy aged 42, and though the methodology involved a hospital-based procedure (after bowel prep), my specialist did say to me that if everyone had a colonoscopy at 40 (maybe as a birthday present) then the rates of bowel cancer would be so incredibly low that it seems to me like I should recommend you think about this as well.

After all, prevention is better than cure. And personalised prevention is even better.

9

The last thylacine

The future of wildlife

'The truth is: the natural world is changing. And we are totally dependent on that world. It provides our food, water and air. It is the most precious thing we have and we need to defend it.'
– Sir David Attenborough

The novel *Jurassic Park* is the allegory I use to explain de-extinction. In that masterpiece by the late, great Michael Crichton, scientists recreated dinosaurs, using other species to fill the gaps in the DNA they had managed to extract from the preserved guts of bloodsucking insects that were around at the same time as dinosaurs. In reality, any DNA extracted from such a sample is likely to be contaminated, and as such, dino DNA continues to elude us. Nevertheless, there are species closer to our current time that could do with being brought back first to help us re-establish carbon cycles in the arctic and carbon sequestration, as well as ecological niches that were keystone species in their time, such as the woolly mammoth, the passenger pigeon and the dodo. Interestingly, the only apex predator we are even close to bringing back from the dead is Australia's thylacine, otherwise known as the 'Tasmanian tiger' (and, even then, the size of the genetic pool of samples may be too small).

The last captive thylacine, often referred to as 'Benjamin', lived at Hobart Zoo until its death on the night of 6 September 1936. Benjamin features in the last known footage of a living thylacine: 45 seconds of black-and-white video, taken in 1933 by naturalist David Fleay, showing the animal in its enclosure. In 2021, a digitally coloured 80-second clip of the footage was released by the National Film and Sound Archive of Australia to mark National Threatened Species Day. Benjamin is sadly believed to have died as the result of neglect. Locked out of its sheltered sleeping quarters, it was exposed to a rare occurrence of extreme Tasmanian weather: extreme heat during the day and freezing temperatures at night. His body was dumped in a garbage bin the next day; it wasn't deemed suitable as a museum exhibit. Official protection of the species by the Tasmanian government was introduced on 10 July 1936, 59 days before Benjamin died. The death was not reported in the media at the time, and the zoo expected that it would soon find a replacement. This, of course, never happened. Benjamin was the last thylacine.

Australian scientists have mapped the genome of a thylacine preserved in ethanol, but a map is not the same as a functioning genome, so the next stage of research is to look at genome functionality while researchers wait for the time and tech needed to reproduce real, genetic Tasmanian tigers. And that time is approaching fast, especially with the advent of CRISPR technology. CRISPR is a genetic engineering technique by which the genomes of living organisms can be cut at a desired location, allowing existing genes to be removed and new ones added. It can be used in the creation of new medicines, agricultural products and genetically modified organisms, or as a means of controlling pathogens and pests. It also has possibilities in the treatment of inherited genetic diseases. The Tasmanian tiger has no extended family to speak of. It is a marsupial, not a dog or a wolf, so crossbreeding for some hybrid vigour would not work. The thylacine therefore remains elusive for now, out of sight but not out of mind.

But why would we want to play at being God and bring back animals that have gone extinct? Well, what were we doing when they were made extinct by humans in the first place? Rewilding could well restore wild places for biodiversity, and also for the functions the ecosystems provide.

Woolly mammoths, coral and climate change

Bioscience firm Colossal has secured millions of dollars to bring back the woolly mammoth. Researchers say the mammoth shares 96 per cent of its DNA with the Asian elephant, so it is likely that a new hybrid species will be created – the mammophant – and it will be as close to the real thing as possible. This de-extinct woolly mammoth will then be released on the Arctic grasslands to help restore that ecosystem and fight climate change through restoration of the carbon cycle, sequestering carbon via their feces at the same time. It sounds like a win-win to me. At least woolly mammoths aren't carnivores or dinosaurs, so the public attitude towards them is universally positive. Plus, the *Ice Age* movies help.

The Lazarus Project in Australia successfully worked with the gastric-brooding frog some 30 years after it was declared extinct. Embryos were created in the lab, and the scientific team is very grateful for the intact tissue that was somehow stashed at the back of a university freezer. Does this amount to cloning or rewriting DNA using CRISPR technology? People are paying some $50,000 to have their pets cloned in parts of the world that allow it, so how much should we pay to see extinct animals re-established in the wild?

On the Great Barrier Reef, accelerated evolution has been employed; however, it is not without its critics. Corals that are particularly tolerant to changes in conditions such as temperature, pH, nutrient levels or even salinity are grown in tanks and then on coral tables, and then planted in coral reef systems to grow alongside native species and help prevent the collapse of the coral systems and biodiversity. Coral is a really complex animal that on shallow tropical reefs operates in a state of ecological mutualism (where two or more species interact with benefits to both species) with algae. The algae uses photosynthesis to produce food for the coral, and the coral provides a clean, safe space for the algae.

There are many mysteries surrounding coral biology. One question fascinating scientists is how coral knows when to spawn. Over time, corals have learned about the moon stages and evolved to time their spawning to a certain number of days after the full moon. I still find the timings and rhythms of Mother Nature to be one of the most enthralling secrets of science that we are yet to fully grasp. Even in recent times, the reef has revealed some more secrets. A new reef was mapped by the *R/V Falkor*, a research vessel operated by the Schmidt Ocean Institute (SOI). The SOI and other research teams are now able to live stream videos of their work to the public. If you're ever needing a break, a trip into the deep ocean via the SOI's remotely operated vehicle, the SuBastian, is a great way to spend ten minutes marvelling at places we will never be able to go to.

Following the wonderful work of Dr Christina Kellogg of the U.S. Geological Survey will lead you to the hunt for 'super corals'. These deep-sea corals are the great-grandparents of the shallow corals you'd be familiar with. They live some 200 metres down in the dark (so there are no algae partnerships there) and have a tolerance to temperature and pH that our shallow-water corals may need in order for our tropical reefs to survive the effects of the climate emergency, and the damage being caused directly by humans through agricultural runoff and upwelling. We need to find where these super corals and deep-sea solutions are hiding. When deep-sea coral research teams applied standard models, they knew from all the data they had from coral research that they had completely missed entire deep-sea coral reef systems. The models were not adequate, to say the least. We have a huge imperative to map the ocean floor as soon as possible, so we can protect what is there.

A map of the ocean floor is planned to be completed by 2030, and there is a major international push for this via the Nippon Foundation-GEBCO Seabed 2030 campaign. This campaign has been formally endorsed by the UN as part of the Decade of Ocean Science for Sustainable Development (2021 to 2030). The timing for this is perfect, because this mapping is becoming more and more necessary in the face of seabed mining and other destructive practices such as trawling, which destroys habitats we can't see.

One way to create an ecological map is through the use of a technique called 'environmental DNA', or 'eDNA'. It's a whole-genome analysis technique from the entire ecosystem, called 'community composition profiling' – that is, working out what species are in the sample, and then calculating the diversity of those species to determine ecosystem status. The eDNA sample can come from water, soil, air or even snow – even the tracks of endangered animals in the snow – and the point is to try to collect a more complete picture than just an individual species' behaviour.

eDNA is a great way to get a pulse check on the genetic potential of an area; but, as with health, we should never have just one metric that we rely upon for conservation. eDNA doesn't tell us much about an ecosystem's function. The other issue with solely relying on eDNA is that it is not quantitative; though some have tried to extrapolate and quantify it, it is usually well correlated in lab experiments but not in the natural world. We aren't there yet.

Economics

How do we conserve ecosystem function, apex predators and formerly extinct keystone species while facing the climate emergency and fighting against humanity's war on biodiversity? One approach is with economics. All ecosystem services performed by Mother Nature have an intrinsic financial value to humans. This is based on their 'benefit transfer values' – what the services would be worth if we had to pay for them. While I can't give you a direct value of, say, a litre of river water in Scotland (because we don't trade in river water), I can give you the value of the local salmon or whisky industry through traditional economics, and we can therefore translate that value to the ecosystem's service of providing clear, clean water for the distilleries and a healthy habitat to support the life cycle of the salmon.

Having worked in a consultancy as an environmental economist for a couple of projects around the EU's Water Framework Directive (WFD), I can tell you how interesting it was to calculate the impacts on Mother Nature. The WFD had directives around controlled chemicals, beach water quality and even shellfish water quality. It was also the first piece of environmental legislation to have economic levers and breaks built in. There was a cost-benefit analysis for restoring or protecting the last bastions of some endangered species, such as the freshwater pearl mussel in Ireland. Once we have a digital twin of planet Earth, we can then look at ecosystem functions on a global scale and start valuing natural assets and resources. This would provide a huge source of income for countries that are currently harbouring huge swathes of undeveloped, wild landscapes and natural capital. What value do you put on a giant panda, a koala or the Murray-Darling Basin? Now extrapolate that globally and you'll find we're talking about a lot of money.

Rewilding

We talk a lot about restoring habitats and recreating ecosystems, but because the climate crisis has already wrought change on a planetary scale, it is impossible to rewind time and re-establish environments that would have existed before the first industrial revolution, or before the human population exploded. We are stuck in a space that can be described by niche theory. Let me explain.

There are two niches to think of here: the fundamental niche and the realised niche. The fundamental niche effectively means all the different conditions organisms require in order to exist. In conservation, this may mean temperatures, food and water access, competition – basically everything that is a basic need to live. The realised niche is the one we choose to live in, or that is chosen for us. For example, some animals can cope with a lot of change, and they adapt. A good example of this is wild animals that adapt well to urban life, such as raccoons in the USA or foxes in London. Effectively, their new living conditions still provide all they need to survive, even if it is not what their species' optimum conditions are.

When we talk about rewilding, we are talking about re-establishing some form of ecosystem function that has been lost. For example, people wonder how the reintroduction of wolves into Scotland might help the population of freshwater pearl mussels. The answer lies in the balance of the niches: wolves eat the deer, which reduces the deer population, so fewer deer trample the riverbanks, reducing the amount of sediment falling into the river. This means that the river may return to a more pristine habitat, therefore producing more food for salmon and allowing the salmon population to grow. The freshwater pearl mussels, whose lifecycle is attached to that of the salmon, can then also start recovering, as their niche is also restored to a survivable level.

We have extensively monitored these environments through remote sensing (RS) and earth observation (EO), using satellites such as NASA's Landsat (which you've probably heard of). This means we can follow our environmental progress (or lack thereof) by measuring change on a landscape scale over decades of data points. These satellites can track the sediment loads in rivers and estuaries, the oxygen and algae levels, and also the soil carbon measurements. We can use satellites to track fires, floods and other extreme weather events, as well as for infrastructure impact assessments and town planning. The best use of this data starts with asking the right questions about the ecosystem you are reviving.

The ecosystem services of this planet provide the very oxygen we breathe, but funding for conservation and environmental protection is lacking. There are things that are so obviously in need of funding, I am agog that we have no billionaires rushing towards these goals. For example, whales are carbon sinks (they absorb more carbon from the atmosphere than they release) – they literally carry the carbon up and down and across the ocean levels, fertilising low-nutrient areas with their poo, and eventually their bodies drop to the ocean floor and take their carbon with them. The carbon sink value for whales is somewhere around US$1 trillion, and yet we don't see any whale-breeding programs happening. These creatures could actually save us from the climate emergency!

Our bodies and our brains love to be around natural wilderness. There's even a term for it: forest bathing. We need green spaces and wild coastlines, and yet we seem to be fighting a war against the natural world – a war we are not going to win. The encroachment of humans into the forests and wild spaces of the planet has created – and will continue to create – global pandemics. We need to listen more to the natural world, create a digital twin of the ecosystem, and question and analyse the data at all levels and resolutions. By protecting Mother Nature, we may actually save ourselves.

Get back in the driver's seat

One of the main issues with conservation programs is that the non-government organisations (NGOs) that run them have to raise funds, which is incredibly competitive. People like to donate to save large, majestic creatures such as whales, or cute and fluffy animals like the giant panda or koala. Clever conservation organisations know this means that they have to translate the protection of keystone species into the protection of the ecosystem that exists to support the cuddlier creatures. Little funding goes to the conservation of a rare soil fungus.

This is where you can come in and assist active conservation. Next time you need to buy a birthday present for someone who has everything, why not contribute a koala or a barramundi to them by supporting a grassroots conservation charity? Check out First Nations conservation charities wherever you live. Some 80 per cent of wild places are under the control and management of Indigenous peoples, and they should be supported economically and socially for their work. If you're not sure or have no easy options near you, then check out well-known and globally vocal options such as The Nature Conservancy, or maybe throw some money at your local university's research and teaching. We all need Mother Nature more now than ever before; it is time to put our money where our hearts, minds and souls are.

10

Vacations, staycations, spacecations

The future of tourism

'Once a year, go someplace you have never been before.'
– The 14th Dalai Lama

In an increasingly connected and globalised world, the future of travel and holidaying has been diversifying both towards and away from technology. In terms of global travel, we have more choice now than ever before; just as we saw the economic surge in the roaring 1920s post-pandemic, so we are highly likely to see a revenge-spend economic boom in travel in the post-plague 2020s.

Travel has always been synonymous with glamour, financial ease and a spark for something different. I remember when I was growing up in the UK, the great TV presenter Jill Dando used to present the BBC's *Holiday* program. I thought she had the best job in the whole world, and I always wanted to go and visit where she had just been. Our 1980s wooden Pye TV was a gateway to the world for me. However, as a teenager working at Boots the chemist and Domino's Pizza in my hometown of Nuneaton (as far away from the sea as one can get in England) saving every penny for university, the tropical beaches of the Caribbean may well have been on another planet as far as the chances of me getting there were concerned.

When I hit 18 years of age, I decided I wanted to have my birthday in a different country each year, if I could make it. I managed birthdays on a winery tour in Cape Town after my gap year to Zambia (19), at a bar in Mozambique eating prawns the size of kittens and drinking cold Castle Lager while watching a South African international rugby match (20), and at the Egyptian Red Sea resort of Dahab (21). Then I started a PhD and didn't have any money for travel at all, though I managed to time a conference in Mexico for one birthday (25) and visited the Yucatan temples, beaches and rarely seen cenotes and national parks (Río Lagartos and its flamingos being a particular highlight). Living in the northern hemisphere with a birthday in August made it easier, because it is during the school and university summer holidays. In the southern hemisphere, it is midwinter and not a major holiday time at all, unless you count Christmas in July!

Travel has been the making of me in many respects, and I've been really missing it – not just because of the pandemic, but because when you're an expat, you end up spending all of your travel time on family visits rather than personal adventures. One of the best reasons for travelling when I was at university was because I could. I relished the freedom – life before mobile phones meant a level of independence we have lost in recent years. Any travel across Africa as a single woman was exhilarating and terrifying at the same time, but I learned the tips and tricks to keep safe, and I was looked after by strangers all over the world, for which I will forever be grateful.

Travel without digital help seems like something that died in the 20th century, and I certainly covet the social media streams of some wonderful travel influencers and celebrities. The world seems smaller, but I am more desperate than ever to start ticking new countries off my list – and I feel like I am running out of time. 'The Travellers'

Century Club' started in 1954 for people who had been to more than 100 countries, but I am not quite halfway there yet. I am just trying to work out how I get to all 193 countries before I die, though maybe not Russia or North Korea any time soon.

Holidaying purely for pleasure is not really fashionable these days – there has to be something to 'do' when you are there. 'Voluntourism' is on the rise, in which people donate a day or so of their holiday volunteering with a local social initiative, even if it is just to speak English with local students. Climbing Mount Kilimanjaro in 2013 was a holiday highlight for me – I organised the trip with my friends to raise money for a British cancer charity and a local Tanzanian research project I was connected to, which was looking at the naturally elevated fluoride levels in the groundwater that were causing disabilities in local children. Kilimanjaro is a life-affirming experience, but the cathedral-like, sky-blue glaciers at the top of the mountain will likely be permanently gone in the next 10 years. My now husband and I did it together, and as a Western Australian it was the first time he had seen snow. We carried a satellite comms GPS tracker with us on the journey so we could ping our families and friends who were tracking our progress, and there was cell phone signal on the mountain.

The COVID-19 pandemic has accelerated or highlighted some changes in consumer behaviour that were already slowly gaining traction. We have seen the rise and rise of the 'staycation', for example. But despite these efforts, the tourism and events industries remain some of the hardest hit, and also paradoxically the ones that will have to get ready to ramp up quickly as the demand to see new people and places will skyrocket (literally) as soon as we are able to travel around the globe again. At the time of writing, so many of our planes in Australia are still in mothballs as the airport hangars of the world are full, and we are in a queue to access the equipment to get those babies back in the air again. It's a frustrating hiccup in the post-pandemic recovery plan.

Why do we need vacations?

Hiring, firing and replacing people costs money, and so there are whole sections of the HR world dedicated to keeping us fit, healthy and happy at work. Preventing fatigue and reducing the risk of workplace health and safety incidents involves keeping people away from the risks, such as through workplace ergonomic assessments.

One principle that became popular in recent years is microbreaks. It sounds good in theory – and I'm sure the people who sell the software have evidence to support it – but there is nothing more frustrating than your computer locking itself when you're just getting into the flow of working on something. It takes a minimum of 10 minutes to get back on task when you have been disrupted, which is why some days you might feel like you've achieved nothing despite being chained to your desk all day. During these microbreaks, you're supposed to move around, get a glass of water, do some stretches or maybe even take a quick walk – but watching the nanny-state software cut access to the document I was working on is the most frustrating game. When I tried it, I stared at my screen and even grabbed a notebook and pen to write down what I was just about to type before I lost the gist of it.

It made me wonder whether these software-based behaviour controls are really helping us rest at all, and whether we can actually recover from our addiction to outputs, timers, social media, calendars and other digital stimulation at all times by any method other than a complete removal from them – a digital detox. I never thought I would be 'managed' by machines, so why does it sometimes feel like the computer is in charge of my day through the endless onslaught of emails and reminders, rather than me using the computer as a tool? Microbreaks just don't do it for me.

When I was pregnant, I tried to keep working up to 38 weeks. As a self-employed person, I didn't take any break of significance after having either of my babies. During those times, I made some bad business decisions and entered into business relationships with people I wouldn't ordinarily seek out. How many of us are making wrong decisions at work or in our private lives purely because we are operating at subpar levels? I remember there being talk of people in corporate jobs having wearable technology deployed to watch for triggers of poor mental health, and I realised then that I have probably been unwisely operating without suitably long periods of rest for a decade now. It's all finally starting to catch up with me, and I know I am not alone in feeling this. Are you OK?

This database of behaviours we have created around our own health and wellbeing has been recognised as a growth area, especially for start-ups created and curated by wellness advocates. Other than running away to a retreat at Gwinganna (yes, please sign me up for a week), there are other speciality holidays that have embraced the removal of all technologies, taking you back to basics (sometimes without even electricity) to try to restart your natural circadian rhythms. Most of you reading this book are likely to have grown up with light available at the flick of a switch, without actually wondering what constant lighting has done to you at a basic biological level. I am a master of the all-nighter and wrote this book's original manuscript mostly between 8 p.m. and 1 a.m., because I needed all of the family asleep and not interrupting me if I really wanted to crack on with writing in a coherent manner.

The closest I have come to getting back to my circadian rhythms is the few times my husband and I have tried camping (with all the comforts we can muster). You end up listening to birds you have never heard before, and seeing the stars in a way your brain can't remember ever having seen them before. You smell the earth and all the microbes within, and maybe even hear the sound of the ocean. Is this what we seek when we travel – to find a place our conscious brain has forgotten about?

Post-pandemic travel

Two areas of the travel ecosystem that have been hit particularly hard by the pandemic are cruise ships and the aviation industry. Cruise ships are now being designed to have intensive care unit (ICU) capability on board and more medically trained staff, as well as passive health monitoring, such as temperature checkers on certain parts of the boat. They aren't quite going to the lengths of wearable technology that some sovereign states and quarantine centres have come up with, but the death of the buffet was not a bad thing (although unfortunately it seems to be a temporary death). You might have seen the videos circulating at the beginning of the pandemic showing how we all transmit viruses – and other things, such as fecal matter – to each other by means as simple as a buffet spoon. Very nice.

At the time of writing, the airline industry is preparing for a resurgence, and in the Asia-Pacific Region we are having to plan for pilot shortages again. This then raises the question of how technology can be used to make up for human deficits. How would you feel about getting on a plane that had no pilot on board? How about if the post you sent was flown on an uncrewed plane? FedEx has already started applying for and testing the assurance and insurance regulatory regimes for cargo planes with no humans on board. It certainly makes sense, given there is a limit on the number of people who can fly the planes, to find alternative methods that still keep humans safe.

The commercial space industry had some hiccups in recent years, and then major successes in 2021, with Elon Musk, Jeff Bezos and Sir Richard Branson achieving their first commercial space flights. The best thing to come from this commercial space race for me was when I saw 82-year-old former astronaut Mary Wallace 'Wally' Funk finally get off-planet. With this 2021 voyage, Funk became the oldest person to visit space at the time. She had trained to be an astronaut in 1961 as part of the Mercury 13 project, but was excluded from spaceflight at the time due to being a woman.

Twenty years ago, the commercial, passenger-carrying space industry was a zero-dollar game. Now it will become a significant player in the estimated US$1.4 trillion space industry. We'll be able to get a ticket to a weekender on the Moon in just five to ten years. Some say that the XPRIZE's 2004 space flight competition sparked the whole commercial space tourism industry.

The Moon has always been touted as a potential launch base for runs to Mars. But would you want to spend approximately five years of your life travelling through space? Another option will be to sample what it might feel like through a simulated version. For decades, scientists have been locked in domes or isolated in the Arctic to simulate what a small group of dedicated space travellers might have to put up with should they spend five years or so travelling and working together on Mars. With the advent of full-body haptic VR suits, it's possible robots on Mars will be programmed by people on Earth doing the work in replica environments in large warehouses – avatars, if you will. You could contribute to the establishment of a human settlement on Mars by doing a bit of light maintenance or engineering that is then conveyed to a robot on the surface of another planet. These possibilities remind me of the very hippie 1972 film *Silent Running*, which featured biodomes floating in space containing different climates, owned by a corporation that wanted them destroyed because they no longer fitted the business model.

Would you take a vacation in a biodome? How about testing your corporate colleagues or team for a week locked in an imaginary research station together? I can't see myself getting to Mars, but I can see my sons having that option in 20 years' time. I'll just go and jump around on the Moon and have an astronaut martini with them to celebrate my 60th instead, please.

What's next for travel?

The aviation industry gets a lot of flak for its carbon footprint and climate change. We can choose ethical and carbon-free travel, and I personally carbon offset wherever I can. ESG investing will affect the tourism industry as high-value, eco-friendly, socio-friendly and authentic travel experiences become more in demand.

It's easier than ever to completely remove the spontaneity of travel plans. We can pre-screen reviews of hotels and destinations, pre-book our restaurants and hire a local guide to assist us. But will there still be space for those of us who loved to travel alone (without mobile phones), waking up in the tent at night as a monkey pinches our toes through the material thinking they are food (banana toes?), or making friends with women on buses and holding their babies or chickens while ignoring the offers of alcohol from the overly friendly men on board? I loved travelling in Africa on my own as a teenager in the 1990s and created some great memories (and I'm sure worried my mum, too, as an unintended consequence). Is this something that will be lost to future generations, who will plan their gap years in minute detail? Or will we see a resurgence in the kinds of unplanned travel experiences my generation grew up with? Will extreme vacations become the next unplanned gap year? Some people's idea of a great vacation is visiting active war zones (yes, really). Others are making and breaking world records in extreme travel, such as the 'Longest Barefoot Journey' (2080.14 kilometres) and 'Fastest Time to Travel to All Seven Continents' (three days, 14 hours, 46 minutes and 48 seconds).

Middle Eastern countries such as United Arab Emirates and Saudi Arabia are spending a lot of money on building entire tourism-centric spaces, convention and exhibition centres, and even new cities, going so far as to build islands and terraform their coasts to cater to the high-value super rich. It makes sense that the oil-rich countries of the 20th century would need to diversify their economies as we look to decarbonise and no longer need their hydrocarbons. Tourism has a significant role to play in the economic viability of the Gulf States and could catalyse social change – especially around women's rights – in some countries that wish to attract Westerners and offer holiday value. Saudi Arabia is already hiring Instagram influencers to visit the country and create attractive tourism content ready for when the world starts travelling again at scale.

The resurgence of global travel post-pandemic will see digital health assets being attached to biometric chips inside smart passports. I have had a vaccine passport for over 25 years, since my preparations for my travels in Africa in the late 1990s included plans for a TAZARA train ride from Zambia to Tanzania. I still have the little yellow WHO Yellow Fever vaccination record book that I took with me and was very careful not to lose. There was a medical team that moved along the train as you crossed the border, and if you couldn't show you'd had a Yellow Fever vaccination, they would stick a needle in you right then and there. I found it ironic that Tanzania, the only country in the world where I have ever had to prove I've had a vaccination in order to enter, years later had a really strong anti-COVID-vaccination sentiment. The idea of vaccine records and quarantine stations is not a new concept, so I really don't understand what a lot of the fuss is about around the latest COVID-19 vaccination records we will need to carry. Just let me on that plane without hotel quarantine being threatened ever again, please.

I relocated to Australia in 2010, in the wake of the Global Financial Crisis, and thought I'd give Australia a couple of years to work on interesting projects and then head back to the UK. But that didn't happen, and I think I can safely say that staying in Australia, getting married, having babies and finally becoming an Australian citizen has been the making of my thirties and forties (thus far). I travelled for work to the opposite side of the planet to try something new. In the future, this is likely to be more common, as people recognise the global opportunities for their careers and our global hyperconnectivity. Bring on the direct Brisbane-to-London flights – what a massive game changer to be able to do more direct flights and fewer hub-and-spoke models. (Is there anything worse than being jet-lagged and needing to wait in transit somewhere for 11 hours to change planes?) And after the flight lands, no digital VR experience can get close to replacing the hugs of loved ones meeting you at arrivals.

Get back in the driver's seat

As the great Madonna song says, 'Holiday! Celebrate!' Maybe it is with the gusto of a 1980s pop song that we need to remember why we loved to travel pre-pandemic. We live on such a beautiful and amazing blue marble that has overwhelming opportunity for travel, but we need to be mindful of the costs to the local area and people if we don't travel carefully and tread lightly wherever we visit. I'm at the age now where I don't have the desire to repeat any holiday twice. I have never understood those that just go back to the same place year after year – how monotonous and boring. There are too many things I haven't seen and experiences to be had; I must see more!

Get one of those maps where you can scratch off the countries that you have visited. Get inspired with documentaries and write lists of things you'd like to do. Carbon offset and check the credentials of any tour companies you use. Choose locally owned where you can. Donate to charities in the country that you visit. Maybe work with the local university for a day and speak English with the students. There are so many small but powerful things we can do to be better travellers. Be a traveller, not a tourist. The world is your oyster.

11

Your 100th birthday party

The future of old age

Ramírez: 'The Kurgan. He is the strongest of all the
immortals. He's the perfect warrior. If he wins The Prize,
mortal man would suffer an eternity of darkness.'
Connor MacLeod: 'How do you fight such a savage?'
Ramírez: 'With heart, faith and steel. In the end there
can be only one.'
– Highlander

The 'singularity' is the hypothetical moment in time when humans create technology that is more intelligent than we are. For some, it's a concept of immortality, of our neural connections living forever in the cloud; for others, it's a pathway to insanity, for we are biological creatures by nature and design. Whichever it is, there are enough people chasing this dream that in a few years, we may view existence, mortality and identity through a completely different set of moral, ethical and technical lenses. These efforts are truly exponential, and when you add the advances in quantum computing and data storage, archiving and access, we are likely to see some large leaps very quickly. We have to prepare for a new class of society – that of the electronic 'zombie', dead but ever-living in someone else's cyberspace. What can it mean to be conscious in such a state?

In every story I have ever read or seen about living forever, from *Eternals* to vampire stories to *Highlander*, there is always a price to pay for immortality, and the biggest price is that of losing who you were to begin with. Consider the thought experiment known as the Ship of Theseus. It is said that a ship, sailed in battle by the hero Theseus, was kept as a museum piece. As the years passed, some of the ship's components began to rot and were replaced. After many years, every part of the ship had been replaced. Is the 'restored' ship still the same object as the original?

If it is, then imagine the removed pieces were stored and, eventually, technology was developed that cured the rot and enabled the pieces to be reassembled into a ship. Is this 'reconstructed' ship now the original ship? If it is, then how can the restored ship also be the original ship?

The concept of four dimensions (4D) is touted as a potential answer to this riddle, with time being added as part of the identity of the ship. Therefore, the same object can be different over time, but still remain true to its central identity.

Thoughts about ageing

I suppose the same can be said of humans. We are dynamic, in a constant state of flux, and change and grow according to multiple sensory inputs and events.

As children, we start with incredibly messy neural networks, and over time we prune them through experience. Neuroplasticity is a burgeoning area of neuroscience, and one that will likely dominate questions of the singularity and even our presence in the metaverse. Could we 'edit out' our trauma, or at least the conscious remembering of it, without affecting who we are at our very core? There are also some annoying earworm pop songs I wouldn't mind pruning off my neural networks ('Baby Shark', anyone?). A favourite movie of mine from the turn of the century is *Eternal Sunshine of the Spotless Mind*, which asks: if we could delete unpleasant memories, should we?

In the Western world, we are facing the challenges brought about by an ageing population. An aphorism that pops into my head from time to time goes something like this: 'It's not the amount of years in your life that count, but rather the amount of life in your years'. I don't think many of us would choose to get to our 100th birthday if we had to spend the last 20 years suffering from severe diseases such as dementia, with no quality of life or sense of identity or control over who we were. It is quality as well as quantity that truly matters to me.

Consciousness as a concept is being challenged and investigated in the frameworks of new and emerging technologies. After my grandparents passed away and I considered all the descendants who exist and will exist, I came to a self-soothing consolation that we never really die – we just get diluted. If we have children, or our siblings do, then our DNA exists in a different form, mixed with someone else's, whose DNA also now exists in a diluted form. I take solace in this, especially when thinking about those who I've lost along the way. We are all ultimately part of the same gene pool, and we have all experienced the same journey; and if we go back far enough, we are all from the same place. What a wondrous and unique thing it is to be Homo sapiens, the sub-species that outlived all other humans.

But why do we even have the biological capability to live as long as we do? Other species – even those genetically similar to ours, such as gorillas – don't live beyond 35 to 40 years in the wild, becoming sexually mature after 10 years old and then having a baby about every 4 years. Effectively we, like gorillas, are designed to live to be grandparents. Imagine if humans today only lived to 40 years old (as was the case a couple of millennia ago)? How would that change how we perceive ageing? The Bible defines the human lifespan as 'threescore years and ten' (that's 70 years), but not everyone is lucky enough to get there.

Plus, the COVID-19 pandemic has actually caused life expectancy to decrease in developed economies such as the USA and Europe.

The future of ageing

At some point this century, we may be able to upload our physical selves digitally into a place beyond the confines of biology. Will we have to work when we are dead, in order to continue to live? Will there be a countdown clock we constantly have to top up? Could we grow bodies and download a copy of ourselves every now and then; or, once uploaded, have we evolved away from being human? Are we going to go insane without a body, or will we find new ways to stimulate the senses and continue to get wiser through life's experiences? Could we edit ourselves and choose to change how we feel about things, self-censor, self-create, and lose ourselves?

As a futurist, I find the best way to digest all the scary what-ifs around potentially horrendous new and emerging capabilities is to imagine that they already exist. Imagine this make-believe world is similar to the movie *Vanilla Sky*, where we can live inside dreams while our body is in the deep freeze ready to be cloned or regrown (a veritable Ship of Theseus), and our brain circuitry, likely bionic now, will be ready to download the artefacts we uploaded and to reconstruct us. The first question is who is going to pay for all this technology, and is it something that is equitable for all of society?

Then there is the question of the right to die. Maybe you don't want to become part of a permanent collective; maybe – and some might call you a Luddite for this – you actually still see life through a 21st-century philosophy and have the audacity to consider that death is actually a part of life, and without it life is, in fact, rendered futile. Immortality is a lie, and to work in a digital form for hundreds of years just to keep paying to live is a circuitous level of hell that

Dante hadn't even considered – slavery to the machine of continual existence. Someone will be making money out of digital immortality, too. Who owns the copies of you? Are there people who are copying parts of you to add to themselves, or are you trademarked, patented, registered and protected by law? Would we still have consent?

There will need to be special insurance for you, and for each of the ways in which you may continue to exist. Health insurance for your new body, upload and download insurance, accidental recovery from the hard drive insurance, the right to delete all copies insurance, the ability to be set to dream mode and disappear as in some form of *Inception*. Perhaps when we can all exist in perpetuity and there are 20 billion of us in cycles of permanent existence, we will stop having children. We won't be animals anymore. We won't even be human; we will have evolved, but into what I am not certain.

All of that makes me feel a bit queasy. It just goes against everything I cling to as a human in terms of our natural cycles of life, the ability to see things through and enjoy the precious time we have. Life is a gift, made all the more special because it does end. But will people feel the same way in 100 years, or will we be renting bodies as living avatars in the way we hire cars these days? It will be interesting to see the effects on people of deep, immersive VR, the metaverse, and even avatar robots off-world on the Moon and Mars.

The economic policies for an immortal citizenry around birth, life and death would have to be completely rewritten. Your 100th birthday party might mean something else, too. Maybe at some point in the future we won't be allowed to age past 100 in our own bodies, because there just aren't enough resources to go around. This is the Malthusian Trap discussed in the introduction to this book – humans just can't control our population growth, and when we are done optimising every natural system to serve the needs of a species that has grown way past its carrying capacity (our natural population limit), we may have to limit how long people live in order to prevent societal collapse.

I recognise that even now in 2022, some parts of the world already treat the poor, the elderly and the infirm in such a way that they are already cutting access to years of life for people, but imagine if it were legally mandated.

Perhaps you find these concepts challenging because your religion teaches you that eternal life exists. Trying to hang around on planet Earth longer than we are meant to may seem like a futile resistance to the truths you hold dear. No offence is intended, of course, and I actually have had some of the best conversations around this topic with people who are on a different path to me when it comes to religious belief. This just goes to demonstrate what a wonderfully diverse group of individuals we all are, and that in society we can be from very far apart places and still respect each other's differences of opinion. Live and let live.

Another great way to think like a futurist is to ask yourself how you'd feel when it comes to your 100th birthday. Are you excited by the prospect? The very idea of ageing that far might actually scare you. At 42, I don't feel like two lots of 21, and the years are going by faster than ever (at least it feels like they are). If or when I reach 100, I hope I would have created a legacy, and that I would be surrounded by my family and best friends, and that my husband would still be with me, too. I'd like to think that I would have had more laughs than tears, and that my loving children considered their childhoods to be some of the best times of their lives. I would like to think that at 100, I would still have some choice, be able to hold the power of consent, and still know how to make a good decision (and a good martini).

Life expectancy and inequality

Back in 2009, it was estimated that half the babies born that year in the UK would reach 100 years of age. This is quite an impressively scary statistic; it also signals that culturally, physically, technologically and economically, we have to prepare ourselves for living longer lives. Human population growth continues to increase, but now not through birth rates as much as because people are living longer. Global life expectancy in 2010 was 69, and it is expected to increase to 76 by 2050. The Japanese hold the record for the most people over the age of 100, and nearly 90 per cent of them are women.

A whole new economic paradigm is about to come of age, called the 'grey dollar' – the spending power of the retired. This assumes that they have access to pensions and savings. It will be interesting to see what role ageing societies have to play when it comes to our economic systems, which are based on having more people working than on pensions. A tipping point is coming in some Western economies, where the population receiving the pension will be too great to be supported by the working population. In some countries, governments have been raising the pensionable age to try to curb the effects of an ageing population. However, the tide is turning fast, and in places such as Japan it has already hit. Other countries have had political programs for population management, such as restricting the number of children citizens are allowed to have. In some places, there are cultural reasons why a male child is preferable to a family than a female child, which have led to baby girls being aborted, left to die, or murdered in hospital dying rooms. In some of these places, there are now shortages of women, and it's likely this will result in a reduction in birth rates.

Unfortunately, the financial systems of the West do not serve women well when it comes to retirement equity – in every developed nation, women are retiring with significantly less money in the bank than men. Older women – especially women of colour – are at risk of becoming the future face of poverty in places such as the USA, the UK and Australia. Every time I learn more about the gender pay gap and how it is not getting better (and is in some cases worsening), it makes me really angry. Currently the end of August is Equal Pay Day; it marks the point at which the average woman now effectively works the rest of the year for free, because men earn 35 per cent more than them.

I firmly believe that the education of girls around the world and the empowerment of women will save us all. When women and girls are empowered, the whole of their society benefits, from health outcomes all the way through to the economic opportunities. The solution to ensuring we can live in a way we all want to, and for as long as we may hope, lies in solving the issues that affect all women in every country around the world. Educated women have fewer children and often have them later on in their life, when they are more economically established, which leads to better health and education outcomes for their children. It is a win-win, and I really wish we could recognise this at the highest levels of the UN, the World Bank and other financial institutions.

The future of mental health and disease

By the time we get to the singularity and have a computer that can work like a human brain, what are the sideshow pieces of research that could be game changers for humans? I am talking about the nastiest neurodegenerative diseases and issues with brain function, such as mental health. A staggeringly large number of people have some form of neurological issue, and this number is only going to increase as our population ages. My dear aunt passed away a few years ago from an insidious disease called multiple system atrophy (MSA), which had been presumed to be Parkinson's until just before the end. Imagine if the brain chips or mechanisms in place to try to process information like a human, or to upload people's consciousness into, could have been used to stop her disease progressing. Neural pathways that had been burned by the disease could have been replaced with synthetic synapses and memories her body had made – not memories of parties or family, but rather memories of how to hold her head up, or control her walking better, or not fall over (especially on the stairs).

My biggest hope is that we start saving some of these stepping-stone technologies, treatments and therapies from obsolescence by recognising that the end goal has a productive (and profitable) pathway, so no useful aspect of what we create can be dumped on the shelf as an 'orphan' piece of technology. There are so many ways in which long-range, grand ideas about how we will have longer, happier lives can add value to an ecosystem of as-yet-undiscovered technologies, applications and pathways.

The right to life and future planning

When we look at it biologically, we are all related – we all evolved from the same place. 'Love thy neighbour', and protect the human rights of all people on planet Earth, from the babies to the aged – everyone has the same right to life. Although the value of a human life has been established in various countries through case law, I just wish that the economic and social systems we have in place could recognise and support everyone's equal right to life in a more active way.

Our current political systems favour public policy that only looks one or two election cycles into the future, but we need to plan a few generations ahead, especially in the face of climate change and a growing and ageing population. Some people are choosing not to have children at all based solely on the environmental impact and geopolitical issues that we are foreseeing.

We have an obligation to protect the right to life, but we also have an obligation to protect the right to death through fully and legally understood euthanasia. Many countries around the world have battled with the idea of the right to die, and I am a believer that it is a personal choice that can be (and indeed is) ethically and morally made with protections in place for the vulnerable.

In the future, what will happen to your digital and online life – all the places there are copies of you, and information about you and your family? We need to ask some hard questions about the right to delete all copies of your social media, employment, health and other records, and, when faced with things like the metaverse and the singularity, wiping the slate clean of all the information you've given over seems like an impossible task.

There is also debate about what to do with an increasing population in terms of the environmental impacts of death and our dead bodies. Embalming and cremation have direct toxic effects on the environment, but we are also running out of graveyard and cemetery spaces, and the wood used to make coffins can be very expensive and detrimental to forests; ergo, there is a rise in environmentally conscientious funerals. Designer and entrepreneur Jae Rhim Lee gave a fascinating TED Talk about the fungus-laden death suit she designed to help increase the rate of decomposition and reduce the environmental impact of your body after you've left it.

We need to start preparing for a change in the way we talk about and deal with death in the next few decades: our rituals and expectations will continue to change. To consider a future world with no physical graveyards, and permanent digital memories and avatars, is to ponder our very relationship with death and what that means for us culturally, morally and spiritually.

I do like the idea of leaving a piece of me, choreographed and polished, as an avatar to have a good chat with and offer advice to my kids and grandchildren when I am gone. As a tech concept, it is already in the making to do a personal 'PS I love you' message post-death. If you could, what would you say? Any skeletons in the closet need an airing with a time lock so they can't be released for a generation or two? My nannas never saw me become a mother, and I would have liked to have been able to call them to ask about the hard and lonely parts of motherhood. They would have had tips and tricks to share, I am sure. No avatar or AI-backed deepfake could give me the hugs they could – and I wouldn't even want them to try – but maybe they could have told me something I needed to hear: something from the relationship that I had with them, something that they only said to me or knew about me, and maybe they could be a reminder of the ghosts of loved ones past who are holding me up when I think I can't make it on my own.

Get back in the driver's seat

We need to get better at talking about death, and about what we want when we pass away. The most burdensome thing for family members after someone dies is the legal and paperwork aspects of wrapping up someone's estate and making sure that they get the send-off they deserve. It's not morbid to plan for these things – it honours your legacy.

No matter how old you are, please get your will sorted. I have seen so many friends and family dealing with the estates of relatives who either had no will or a will that no longer reflected the relationships they had in their lives. A great place to start is *The Bottom Drawer Book* by Lisa Herbert.

Research funeral homes that offer environmentally sensitive options, such as not using preservatives before cremation, or using sustainable products. You can even choose to rent caskets now and reduce the forestry burden. It's also worth reviewing your insurance policies, particularly life insurance that you may have attached to your private pension fund.

It is such a hard and emotional time when someone dies. The last gift we can give to our loved ones is to have our affairs in order.

12

The war with no dead

The future of geopolitics

'We shall go on to the end, we shall fight in France,
we shall fight on the seas and oceans, we shall fight with
growing confidence and growing strength in the air, we shall
defend our Island, whatever the cost may be, we shall fight
on the beaches, we shall fight on the landing grounds,
we shall fight in the fields and in the streets,
we shall fight in the hills; we shall never surrender...'
– Winston Churchill, former Prime Minister of the UK

I am one-quarter British paratrooper. My paternal grandfather, Gordon, was in the Army Air Corps (red beret) and part of the team that famously made the airborne raid on Bruneval in late February 1942. This daring raid, called Operation Biting, aimed to gain information about the radar systems the Nazis were using in northern France. Until then, the exact purpose and nature of the radar systems was not known. Operation Biting was a massive success, with troops bringing both technical parts and a captured German radar operator back to Britain to assist in the race to stay ahead with radar technology.

Thinking about my plucky ancestor gives me strength (and belligerence) when facing my own challenges. My grandad used to go for extra jumps just for the cash, and was unfortunately seriously injured during the D-Day jump into Caen. He suffered a life-changing head injury, which left him with amnesia for at least a year. Like many of his generation, he never talked about the war – more's the pity.

In the future, there will be fewer and fewer humans making up the troops on the front line and behind enemy lines. This brings me some comfort, but it also increasingly gives me worries about the safety and security of future generations. Warfare is evolving into something less visible to the public, and on many levels there are wars we are already fighting that most people don't hear about on the news. Cyberwarfare has been going on for decades already: worms, Trojans, bugs and bots all infecting our internet, our computers and our data storage systems. This is not traditional warfare, and it doesn't follow the rules of war or the Geneva Conventions.

Sometimes it feels to me that the world is more connected than ever, and yet as individuals we are more disconnected than ever before. It's paradoxical that we can connect as unique people to other unique people, yet as societies we can sit and imbibe news every evening about another fall of government, another use of unethical or illegal weapons of war, another instance of human suffering, and be almost numb to it all. Have we been overexposed to war through traditional news media and new social media? Are we feeling unable to instigate any changes ourselves, and so we sit with empathy fatigue, unable to help, unable to stop it, unable to convince politicians to see our world view? Are we all 'cried out', especially following the emotionally draining pandemic? Have we disconnected ourselves from the idea of going into battle à la D-Day, because now we have robots, drones and AI to take the heat of the war instead?

Robotic warfare

There are so many conversations happening around the world about the ethics of autonomous warfare. The stories being shared about defensive technical capability suggest that we may get to a place in my lifetime where humans no longer jump out of planes to do daring manoeuvres in the dark over occupied territory like my grandad Gordon.

An expression that originates from military use and is now applied to all robotics, drones and AI-backed hardware is that new and emerging technologies are great for the 'dull, dirty and dangerous' jobs, including in the theatres of war. Remotely piloted or automated hardware has been used for dangerous frontline combat work since World War II. The Russians and the Nazis both developed remote-controlled mini tanks and demolition vehicles, and the design of this group of robots hasn't really changed – you can find counterterrorism robots and drones that look a lot like the miniaturised version of tanks developed 80 years ago. However, with the advent of soft robotics and bio-inspired engineering, there are some that are looking very different.

DARPA developed a surveillance drone that could fly in all directions – they had effectively modelled it on a hummingbird. In some movies recently, there have been drones that are the size and form of small insects in order to conduct surveillance inside houses without detection. The capabilities of drones, robotics and other automated or autonomous systems for warfare are diversifying and accelerating, with some also converging.

The major problem with all new technologies is that the legislation and regulation is always lagging behind the actual technology, and so broad points of principle have to be established. For example, there is a growing call for a ban on all killer robots based on existing human rights legislation. Also known as the 'Martens Clause', this legislation could require all automated, autonomous, robotic, emerging technologies to be assessed against the principles of humanity and the public conscience when they aren't already explicitly covered by other parts of human rights conventions. Countries are working collectively to create a ban on fully autonomous killer robots, but this requires open and honest collaboration and relationships, and current geopolitics might make it difficult to trust this process.

Just as one cannot predict the future of the stock market based on past performance, we need to understand the concept of 'survivor bias' in conflict, especially when it comes to autonomous and automated systems. One story that springs to mind is that of the Royal Air Force bombers coming back from bombing raids in World War II, during which they were shot at with anti-aircraft fire. The returning planes were examined to see where the bullet holes were, and someone thought that where they'd been hit should be reinforced. So, heavily reinforced bombers were sent back into the skies, never to return. The truth that was missed was that when planes had been hit where the bullet holes were, they could still fly back, but if they were shot in other areas, they didn't return.

One of the surprise benefits to the Allies from the jump at Bruneval that my grandad did was that the Nazis started reinforcing their radar stations with more ground defences, such as ditches and barbed wire. Ironically, this made them more visible from the air and therefore more easily identifiable as targets for bombs.

When it comes to direct combat and conflict, how will future armies be trained? Will soldiers have exoskeletons? Will they have been picked using AI-backed mass surveillance of primary schools? Could they be bionic or genetically modified? Or all of the above?

Some of the technologies that have been created to try to assist the most innocent of all human life, newborn babies, could be used in a military context. In a world in which premature birth is the greatest cause of death and disability among children under five, there are many people working to help those babies develop more before they enter the real, harsh world. Currently evolving artificial womb biobag technology will make out-of-body gestation a reality. Since 2017, this technology has successfully grown lamb fetuses that were cut from their mother by C-section and then placed in the biobag.

What would this mean to a country that wanted to create mega-armies? It could mean an army of soldiers with no mothers, whose parent is the state itself: *The Truman Show* meets *The Matrix* meets *Never Let Me Go*, with genetic modifications and bionic adaptations. The super soldier of 1980s sci-fi movies could very well be a reality this century.

Cyberwarfare

Some of the most insidious wars are happening online. Cyberwarfare has been happening for decades. You may have experienced this on a personal level if you've ever been hacked on social media or had your identity stolen. The cost of cybercrime to the global economy has been estimated to sit at around US$1 trillion annually, which is approximately 1 per cent of GDP. The most worrying thing about this data, though, is the trend – an increase of 50 per cent between 2018 and 2020.

When I was working a corporate job, I did the basic online training modules around cybersecurity and got the certificates, and I never thought I could be the person to slip up and click the wrong link in an email. Spam has certainly evolved since the Nigerian princes of the early 2000s needed my bank account details to transfer millions. Now some spam emails mimic utility bills and bank accounts really well. I was once sent a test email at work. It looked like a Valentine's card, but I was immediately suspicious. I didn't click on the link, but I hovered my cursor over it and saw the website name, which I then looked up on a search engine. It turned out to be a company that works to test your staff to see if they will click on silly links. I passed the test (thank goodness, because that would have been embarrassing), but really it is now my default not to trust emails from people I don't know, especially with links or attachments.

There is now more spam content on the internet than real content. This means the managing of accounts and expectations on the internet has become more difficult, and therefore more expensive. Some people think that working from home or remotely is the worst thing we could do for the cybersecurity of our personal and work systems, and yet many people had to during the pandemic.

Cyber attacks aren't limited to emails or website viewings; they are also coming straight down the phone. I get a good few 'robot' calls per week, and I never say hello first if I am unsure of the number, because I always want to check if there is someone there. Just like email phishing (the 'ph' stands for 'password harvesting'), voice phishing can vary in terms of sophistication, and can sometimes be threatening or pretend to be from government agencies. Some are automated robotic processes, but other times actual human beings can be making the calls.

Another scam now involves audio deepfakes: people get calls from the voice of a trusted individual asking them to do something. This shows just how far scammers will go and how much money they must be making to keep trying these things.

Cybercrime might seem like it has nothing to do with warfare, but by targeting the most strategic industries, institutions and essential assets, a successful cyber attack could cause more damage to an economy or political party than any direct conflict would. At the start of the COVID-19 pandemic, there were major cyber attacks on the universities and institutions working on the virus, and especially the laboratories working on the UK-based vaccine. We don't have 007, but rather code writers and AI that is data-gathering in foreign countries now. I bet they don't even drink martinis.

Information warfare

Another tactic is the subtle but effective use of information manipulation, aka 'information warfare'. This type of warfare is the digitally evolved version of dropping flyers over enemy lines with demoralising content, or blasting information by stereo across the border, such as occurs between the Koreas. The bad news is that you have already been exposed to information warfare, even before any foreign agency started funding anti-vaccine information in 2020. The core modus operandi of information warfare is to manipulate information that would normally be trusted by targets (people, businesses or agencies) without their knowledge, causing confusion about what is real or not and eroding trust. Information warfare can also be used to test and verify the reach of the nefarious actor's actions, providing them with assurance of their influence and capability, like cartographers' trap streets on maps.

The use of information warfare to spread propaganda, false information and fake news really came to the world's attention during the 2016 US Presidential Election campaign, but that seems to be dwarfed in comparison to the false information being spread across traditional and social media about COVID-19. Science communication is hard at the best of times, and personally I believe if we had better STEM literacy, a lot less of the fakery would get through the filters and fewer people would share it. Unfortunately, it seems that misinformation and manipulation are well accepted by individuals even during a global pandemic, when false words could cause the deaths of people swayed by them.

Electromagnetic warfare

Electromagnetic warfare involves the use of the electromagnetic spectrum to block or destroy communications or radar systems – you may have heard it referred to as 'jamming' someone's signal. It has been around for a really long time, and much research on it has been carried out in the 'communication-dark' areas of the Woomera Range Complex in South Australia.

Combine cyberwarfare, information warfare and electromagnetic warfare, and you create the need for people to prepare countermeasures, which means operations without technologies. Technology is being developed that mimics the navigation mechanism of birds by using the Earth's magnetic field, and even using the stars in the sky for navigation. All military ships have the capability to navigate through traditional means of celestial navigation using the location of the sun at midday, the stars at night, and a piece of old technology called a 'sextant'. When it comes to combating and surviving sophisticated attacks of the 21st century, the best methodologies may be the oldest.

War in space

Wars have been fought all over this planet: on the sea, in the air, on the land, in homes, on the streets and in our social media feeds. But what happens when we leave Earth? Does the international community expect all the human rights and laws of engagement to be replicated off planet? Could we have a war in space? How about in the metaverse – is that still deemed to be planet Earth?

There are certain things we already know: we have robots that can drill samples out of asteroids, and now we have companies set up to mine asteroids for resources such as iron ore. We have a Moon Treaty from 1979 that was never ratified, just as we have a number of nations wanting to establish their own Moon bases. On Mars, a number of exploratory robots are active on different parts of the planet. What could this all build up to?

Planet Earth is a closed-loop system. We have finite resources and a culture of consumption and development that is hard to curtail. It certainly won't take long before the international treaties protecting places such as Antarctica from drilling and mining expire, and certain countries may not sign up to any new restrictions at all. What happens then? Do we go to war to protect Antarctica? On Earth, we have a shortage of helium, an element that is plentiful on the Moon – so will we allow mining there? The Moon affects our ecosystem (it controls our entire ocean ecosystem and a lot more), so changing it significantly through mining could have planetary ramifications.

Then we get to Mars and other international space 'assets'. How will we put humans on Mars? Will there be a desperate land grab, like with the colonists of the past, or will there be bidding wars for contracts to drill, as with the resources industry? Would we go to war on Earth over rights on Mars and all the resources there? As the problem of resource depletion on Earth becomes more real, the idea of extracting valuable elements from asteroids and returning these to Earth for profit, or using space-based resources to build solar-powered satellites and space habitats, becomes more attractive. Hypothetically, water processed from ice could refuel orbiting propellant depots.

Although asteroids and Earth accreted from the same starting materials, Earth's stronger gravity pulled all heavy siderophilic (iron-loving) elements into its core more than four billion years ago. The crust only has such valuable elements as gold, cobalt, iron, manganese and nickel because asteroid impacts infused the depleted crust with them. It is believed that many asteroids contain such important elements. In 2006, the Keck Observatory announced that the binary Jupiter trojan asteroid 617 Patroclus, and possibly many other Jupiter trojans, consist largely of water ice. Similarly, Jupiter-family comets, and possibly near-Earth asteroids that are extinct comets, might also provide water. Ice would satisfy one of the physical needs to enable human expansion into the Solar System.

Climate change and the 'world peace' ideal

Resources, it seems, are one of the things humans are likely to go to war for, and on a planet that is facing a climate emergency it is worth noting that countries have already gone to war over precious natural assets, such as Ethiopia and Eritrea over water supply and access to transboundary rivers. When the first climate change refugees make moves to relocate, and when climate change has significantly negative effects on areas such as food production, pest control (such as locusts) and water access (such as in the Himalayas) that increase the number of people who are disaffected or needing to relocate, we are likely to see increased geopolitical tension. Climate change will further accelerate and augment existing tensions between neighbouring countries over dwindling resources and growing populations. The climate war will not only pit Mother Nature against humans with increasing frequency and ferocity of extreme natural events, but also create new philosophical and political differences and ideologies within and between countries all over the world.

One might take a more positive attitude towards war and peace and hope that, with the betterment of society through technology leading to better communication and more common ground, the global fight against shared problems and the increasing globalisation of society will make wars less frequent and eventually cease to be. I am too much of an optimist even for myself sometimes. Could we ever get to a place of balance where there was enough for everyone and therefore nothing to be fought over? Are the economics of war and defence such that they are too big a machine to stop, or will working together to help Homo sapiens survive create the nirvana we have all been waiting for?

The core desire of supposedly every beauty pageant on the planet is 'world peace'. One might argue that the use of AI for conflict must surely mean we could also use AI for peace? There is a push for #Tech4Good, #DronesSaveLives and #AIforGood on the social media channels of tech evangelists, and the move into Industry 5.0 means 'profit for purpose' is becoming a business validation. Call me a dreamer, but somehow I know that when we work together as a species, there is nothing we can't solve.

Get back in the driver's seat

Cyberwarfare is an increasingly malevolent threat. There are even hypnotic maps tracking cyber threats live – a good one is threatmap. checkpoint.com. A good starter for a dinner-party conversation, perhaps?

There are plenty of things you can do to reduce your risk of cyber attack. It almost goes without saying that you should never give out your details over the phone. If you aren't sure if a call is genuine or not, hang up and call the actual entity they say they are calling from. (This all sounds easy as pie, except I remember trying to educate my grandmother about her mobile phone and text messaging, and her total lack of confidence with technology. It's a worry.)

Make sure you have antivirus software. Set your smart devices so that they complete software updates automatically. Never click on links that come in text messages, emails or DMs from people you don't know; even if you do know the person, text or email them back to check they sent you the message if you see there is a link attached.

Multiple-factor authentication (MFA) is a saving grace for *all* your socials and your web-based emails. Make sure you have rescue emails and verification set up. It's so easily done, only takes a moment and really helps protect you.

Also, next time you're visiting your parents or grandparents, ask them about cybersecurity and set up their verification processes for them.

13

Everybody needs good neighbours

The future of networking

'... Am urged by your propinquity to find
Your person fair, and feel a certain zest
To bear your body's weight upon my breast'
– Edna St. Vincent Millay, 'I, being born a woman
and distressed'

The average community has changed a lot since I was a child. Growing up, we knew everyone closest to our house in the street. We knew where the kids our age lived, where the oldies with the best choccies were, and where the slightly odd people were (and how to avoid them). We sent Christmas cards to our neighbours and helped out if they had a fall. We rescued runaway rabbits and babysat local toddlers. We had paper rounds on our estate, worked in the local shops, and walked to school come rain or shine. We walked people's dogs, fed their cats, and picked up litter from the ground and put it in the bin.

Some of these concepts might seem unfamiliar to you if you're younger than me, because we have experienced a paradoxical shift – we are now more connected to those we know and strangers on the internet than ever before, but many of us under 50 are not likely to know who our physical neighbours are. Before the pandemic-related lockdowns, how often did you say hello to a neighbour, borrow some sugar or coffee, or water someone else's plants?

During the lockdowns, things changed a bit. In the UK, one of my friends made personal contact with an older gentleman on her street. She already knew him well enough to say hello to, but was now grocery shopping for him, cooking food for him, and making sure he got a vaccine as soon as they were available. She became his extended family, because his own family was too far away or unable to help as much as they might have liked.

Social media is wonderfully connective too, and it has allowed us to find old friends, travel buddies and university mates. We are able to contact people on the other side of the world via free video calls. Remember the videophone booth in the movie *2001: A Space Odyssey*? Science fiction has become real and readily available.

But problems have arisen from this immediate connection. My family has had a series of fallings out: defriending, blocking and removing connections because of family politics, perceived slights and full-on arguments. I think that, on the whole, my experience on Facebook has been more negative than positive. Being connected or linked to people who you don't actually want in your life anymore, or even to be reminded of, can perpetuate trauma that might otherwise have been dealt with quietly, healthily, offline. We sometimes get tempted to check on our ex-lovers or former friends, as well as former enemies, bad bosses, evil colleagues and random people we would never normally have connected to. Not to mention the physical issues: people are reporting bad necks, migraine headaches, repetitive strain injury and even needing glasses earlier than normal because they have been spending far too long on social media via our phone use. What on earth are we doing to ourselves?

Where's the connection?

When our addiction to technology starts taking more of our time than our genuine connections with real humans, we need to take stock of ourselves and our behaviours. In London, councils started putting padding on lampposts because people kept walking into them. Some pedestrian crossings have now been developed into 'smart crossings' – with lights, sounds and changes to the texture of the pavement – to stop people from walking onto the road without looking up and checking for traffic. It's as if we are sleepwalking, because our connection to the digital world seems more exciting or safer than the real world and real-world interactions.

The average user of social media spends 2.5 hours on it per day, which adds up to nearly 1000 hours a year. This is shaping our relationships and guiding us to spend time with particular people, some of whom we might never meet in real life. It reminds me of *The Net*, in which Sandra Bullock plays a tech worker who has no friends and has never met her colleagues. Even the FedEx guy isn't sure what she looks like. So, when she gets entangled in some corporate bad guy's mess and they try to pretend that she doesn't exist by replacing her with someone else, her character can't prove who she is! It gave me nightmares.

Theoretical structural chemist August Kekulé was struggling some 150 years ago with how the benzene ring formed and what shape it was. He was increasingly frustrated that in a burgeoning area of chemistry – aromatic compounds – the standards of structure were just not fitting with the empirical formula that was well known. One night he had a dream of an ouroboros (a snake biting its own tail), and this helped him solve the structural conundrum of the benzene ring as a six-carbon structure, with alternating single and double bonds. And so, a new era in aromatic chemistry was born. I have the same dream as Kekulé, except mine is all about the ouroboros of social media, and how we just become increasingly circular in our interactions. It is like we are almost biting our own tail. Like filming ripples on a lake caused by throwing pebbles, and then playing that backwards, we are removing the growth and evolution of our real social interactions as we choose to direct time and energy to our relationships through social media.

In the online world, how do we know what is really real? Are we ever really our true selves behind the screen? You may have heard the term 'catfishing', which means to create a fake or fictional person in order to extort money or get some kick from manipulating people. The sci-fi short *Hashtag* shows an extreme version of this. It depicts a social media influencer with millions of followers who was making someone else a lot of money through sales. People thought she had the perfect life, but in reality she was living in a box made by her own mind. Eventually someone broke her out and, more than any kind of desire to live a real life, she was petrified about losing her fame.

On social networks, aren't we all just catfishing? Could it be said that no-one is genuine on the internet, and therefore we are living separate lives, with online personalities that don't reflect reality? I often wonder about the trolls who send death threats one minute and then go back to their happy, normal family lives the next. It feels like social media is allowing us to have multiple personas, and is normalising this split in behaviours.

Some people would go so far as to say that social media is fuelling a new world order and is, in fact, an incredibly dangerous thing. I think we are at a turning point now, and we need to decide what we want. Breaking up social media monopolies would allow some oxygen back in the room in terms of competition and laws around data sharing. The idea of spending 1000 hours a year – some 41 days – on creating social platform profitability for billionaires seems like a sick waste of human effort. Imagine what you could do with 41 solid days. That's like having another five months' worth of weekends a year. I wrote the first-draft manuscript for this book in 20 days. When we focus, we can get a lot done.

There's also the question of equity. This digital part of our lives is not equally accessible to everyone, so there is a risk we are creating and supporting some kind of new class divide. The digital natives, digital immigrants, digital integrators and the people who have never even touched a smart device all coexist on this planet, but not all are connected. Unequal access means unequal progress in global society.

What are we missing out on?

The element of surprise is something we used to enjoy, particularly when interacting with other real-life humans in the flesh. The accidental, opportunistic, serendipitous moments added joy to our lives. Before dating apps and social media, speed dating reached a peak in the early 21st century, and the funniest stories came from those who did not meet the partner they were hoping for. Isn't that one of the joys of life – to have experiences that are not tailored, calculated or curated directly for our consumption, but rather allow us to have a laugh at life and ourselves?

What are our lives like if we always see what we want to see, and we never see anything outside of that? Our boundaries are unknown places, and if we stop pushing them, our journey is done in many ways. Who knows who you're going to meet at a random coffee shop in a cute village, and where that meeting will take you?

My mother bought me a book for my birthday years ago now, and little did I realise that it was my first ever William Gibson: *Idoru*. It was the first futurist cyberpunk book I read; and thank you, Mum, because I can remember it now. This book was first published in 1996, and Gibson really was a great future thinker because it refers to things we have in our lives right now and that are coming soon. The crux of the story is that someone wants to marry an AI, VR pop star, who is neither human nor in corporeal form. There is an attempt to humanise this AI-based celebrity, which has a personality construct that evolves with each human interaction it can get, and also changes shape and form depending on what her fans want to see.

The idea that humans can be attracted to AI-based personas has come up a lot in recent years, from the movie *Her* to *Blade Runner 2049*, to *Ex Machina* and Steven Spielberg's *AI Artificial Intelligence*. However, we are actually doing ourselves no favours humanising AI or robots. There is a risk in assuming that AI can actually gain EQ (emotional intelligence). AI is actually not particularly proficient at reading the subtle cues of human feelings and emotions. While many mental health first aid apps – such as Woebot (a robot you tell your woes to) – might seem to display empathy, really they are programmed to behave in a certain way.

Sophia the humanoid robot was activated on Valentine's Day, 2016. She was designed to imitate human facial expressions and gestures, and is able to answer certain questions and participate in simple conversations on predefined topics (such as the weather). I sat on a panel a few years ago with a pre-programmed Sophia, who we weren't allowed to talk directly to because she wasn't actually running on a chat-bot basis, just a pre-recorded one. I have to say I was a bit disappointed, really; I was hoping I'd get to interact with a machine-learning superstar, not just a robotic-looking puppet.

Sophia's creator, Hanson Robotics, have now launched 'Grace': a humanoid female nurse robot who will work alongside doctors to give direct patient care. Grace has a number of sensors that collect patient information, including thermal cameras to measure pulse and temperature. Grace will ultimately be a support robot in aged care – an area that needs more people but seemingly has less investment than needed.

Will robots become our friends and supporters when no-one else is there? How will that compare to real human support? Will some people substitute their obligations and delegate their responsibilities when it comes to healthcare and aged care by hiring or buying a robot companion and nurse for their ailing relative? There are positives here for people who don't have family to assist them or whose relationships have broken down (as they do), and for people who are social distancing during times of pandemic management, or when there is a lack of staff due to recruiting or pay issues. However, we need to build checks and balances into the systems in order to protect the vulnerable, because I don't think we have the social OK yet for something like AI or a robot being able to switch off life support or assist with euthanasia.

During a pandemic is a clever time to develop healthcare robotics and automation, that's for sure. Hanson Robotics is planning to start mass-producing and selling the Sophia and the Grace robots in the next few years. Want one? I think our aged-care sector needs a few.

Professional social networking

I've found it interesting to see the resurgence of in-person events and conferences in the post-pandemic landscape. Our choice and diversity of human interaction and community curation were taken away by COVID, and it seems that whenever we can get together at an event now, we are all the happier for it. Perhaps we have actually reached our tolerance level for purely online interactions. We want to meet new people, have accidental networking opportunities, make new connections over work and professional conversations, and sit and have coffee with real people.

There is something to be said for the non-verbal cues and information that are part of how we communicate as human beings, too; these small but significant movements can tell you more about people and how they are feeling than the words coming out of their mouths. When we take time to meet existing online connections in the flesh, make new acquaintances and find mutual friends, new and potentially interesting opportunities can come with it. The more we build social networks, the more we realise we need IRL, human-to-human conversations – not just digital ones.

I think we have all search-engine-screened potential new employees or even first dates before we have met them in real life, and perhaps that is just a digital version of the old ways of gossip. We all just hope that the search engine algorithms show us in a good light. People have been known to get fired, or not even get hired in the first place, because of something posted on a social media page or website. Interestingly, defamation law is now catching up, and recently in Australia a judgment was made that any news corporations sharing comments from social media that may be defamatory are just as responsible for that defamation as the people posting it and the platforms sharing it.

As I have been known to say, your network is your net worth in terms of career and relationship management, and even business or entrepreneurial success. But our networks also inform and shape us. Many studies show we are the products of our communities – we are the sum of the five people we spend the most time with. But what happens if one of those five people is an old school friend from a different life who we don't interact with – it's just that she posts a lot on social media? I don't think we can carry the old-social psychology into the new world. We need to consider tech blinkers, making sure we don't get overloaded or taken off task by weapons of mass distraction. Filtering, blocking, muting – these are all good tools. Use them early and use them frequently.

Get back in the driver's seat

Ultimately, you are in control of everything you do and every action that you take. Legally, morally, neurobiologically, you are the master of your fate, you are the captain of your soul. Life can get complex: people are complex, relationships are complex, and how we interact with and use technology has gotten very complex, very fast since the start of the 21st century. In the movie *The Matrix*, the AI states that it recreated the immersive world as the late 1990s because it was the pinnacle of human civilisation, but I don't necessarily agree. These times between big, obvious, swinging changes can be described as liminal or transitory, but they are also foundational.

Review your life and your relationship with social media. What would happen if you didn't use your devices for a week? How about if you gave yourself just an hour a day on social media – timed, intentional, powerful and active, rather than passive warbling and doomscrolling? Social media will not be disappearing any time soon, so think about the ways you can make it work for you. And follow me on Twitter @DrCatherineBall.

Someone once said to me: send texts or make social media posts as though they could be printed out and posted around your neighbourhood or read out in court. Never post other people's personal details; avoid posting images and information about others without their consent; and when it comes to the algorithms, never, ever tell them your darkest secrets.

14

The roof is on fire

The future of the planet

'But how can I feel the burning of a match,
when I am being held in the arms of the sun?'
– Elizabeth Smart, *By Grand Central Station*
I Sat Down and Wept

Are you feeling overwhelmed and out of control when it comes to the sheer volume of information we are fed about the bad things happening in the world today? These are things we have very little actual personal control over, such as climate change, biodiversity loss, ocean plastic and the dangers facing the natural world, as well as famine and war. Most reasonable people do not want these things in their lives, but the governments we elect continue to perpetuate them, such as by subsidising the fossil fuel industry

The economic impacts of a changed climate are already here and are going to get worse, which will lead to parts of the world becoming completely uninhabitable (and uninsurable). What happens when an entire city is not able to be insured? The financial systems of the world are built on assurance of value and of return, not on tipping points of failure that are one-way doors. Surely, then, there must be some reasoning that the way we have always done things is the antithesis of how they should be done from now on? And yet there appears to be such resistance and pushback from the leadership in government and boardrooms around the world.

Nearly all boards I know have a risk subcommittee, and surely these committees must discuss the climate emergency. Yet, as I mentioned in chapter 2, fewer than 3 per cent of Australian company directors have a STEM background. When the future is at stake, surely we need people who are scientifically literate leading the charge? We also need more people with PhDs – who speak the language of statistics, innovative thinking and problem-solving – on company boards, perhaps mandated through quotas.

Diversity of thinking and innovation are two sides of the same coin. If there is one area in which we need businesses to recognise the imminent issues and deal with some serious innovation or mitigation, it is with regard to the climate emergency.

Eco-anxiety and education

If you have been frustrated by media reporting on the science of climate change, please feel welcome to join the club. There's even a new term to describe our feelings of helplessness with regard to the obfuscation, misdirection, fakery, false commentary and opinion taken as fact: 'eco-anxiety'. A similar term exists in German: *weltschmerz* (a word dating back to 1827 that literally translates as 'world hurt'), which encompasses a genuine ennui and sentimental pessimism about the world. The slow drip of worry has become a tsunami of panic. Young people are starting to make their feelings heard, with some even looking at legal avenues to protect themselves and their futures.

As the adage goes, 'We don't inherit the planet from our parents, we borrow it from our children'; but, like a messed-up library book, we have dog-eared the corners, spilled coffee on it and forgotten to renew it. We have been found wanting, and it is up to us to get our collective act together and make the changes we can while we can.

Where should we start? The first place is actually to get educated via reputable academic sources in climate science, and to start thinking like systems engineers. There has been no fundamental change to climate physics since the year I was born (1979), and so there is no debate to be had around the mechanisms of climate. Despite this, there are still those who would disagree and argue about science that they barely understand, and this says more to me about trust (and the lack thereof) of science and evidence from some elements of society.

'Climate emergency' is a term that evolved from 'climate change'. It began gathering pace and grabbing headlines at the end of 2019 (when it seemed as though half of Australia was on fire). Then the COVID-19 pandemic took over the globe and took all news media channels with it, and no-one wanted to talk about climate change anymore; it was all about the pandemic and getting life 'back to normal'. But what if normal is not like it was before? Adversity is the mother of invention, and this may be a liminal time and a golden opportunity to press the reset button on how the global economy works, and how big business (global) influences politics (local).

Thinking back to Malthus and Solow, who I referenced in the introduction to this book, climate change does still cause friction between people, even friends. It has become so black and white: either you care about climate change, or you don't. There is a grey area: people who do care about stopping bad things happening, but also don't think that we as individuals can make any difference, and so they don't keep the climate front of mind when making purchasing or lifestyle decisions. This is a problem, because real change in the world requires change in human behaviour. A cartoon about this that made me laugh, just because it was so painfully obvious, showed a boardroom full of suits. The person who stood at the front asked the question, 'Who wants change?' and everyone raised their hands. Then they asked, 'Who wants to change?' and no-one raised their hands. Then they asked, 'Who wants to lead the change?' and everyone left the room.

Climate scientists have now worked out that we can't rebalance the increasing temperature by cutting emissions alone; we actually need to remove carbon from the atmosphere or cool the planet in some other way. So, we are no longer looking at just adaptation strategies; we have mitigation and carbon-reduction plans that we need to embark on. Unfortunately, many companies are getting the greenwashing PR machine going. One that made me spit out my tea last week was a soft drink company advertising their bottles as recyclable and pushing for recycling, but fewer than 9 per cent of plastic products that could be recycled actually are. There are stockpiles of plastics waiting for export from Australia and other Western countries to go to overseas recycling facilities. Lots of people get angry about these failings of the 'reduce, reuse, recycle' movement and go so far to say it is merely all talk and no action. I look at the piles of plastic waiting to be recycled and think about all the oil that was drilled out of the ground to produce it, and the carbon footprint attached to it. You can't separate messes like ocean plastic pollution from climate change – they are directly connected.

A technology-led approach to saving planet Earth

Data oceans and AI-backed modelling software are already being used to develop digital twins of our cityscapes. The ultimate goal, in terms of understanding planetary processes, would be to create a digital twin of planet Earth. Some researchers are already attempting this, but it needs a coordinated team all working with the same metadata framework. It would be one heck of a neural network or systems engineering approach to create, store, archive and interrogate a digital interface such as this but, like the human genome project, it will get easier and faster as we get better at it.

The roof is on fire

At university I was always a bit confused as to why environmental science and marine biology were taught as different degrees with very little crossover. It frustrated me at the time, because we all know that there are no hard edges when it comes to Mother Nature. To study rivers is to study estuaries is to study the water cycle, for example. So, maybe we need to teach environmental science via a more systems-led approach. The digitisation of assets such as earth observation, water quality sampling and even drone mapping of coral reefs should all be seen as nodes of a complex network, rather than information silos. Geoengineering as a solution to climate change has been around for a while, and the biggest problem facing geoengineering with regard to climate change is that it is a planetary-wide endeavour and would affect every living thing here.

Some of the carbon-absorbing ideas that have been discussed over the past 15 years and are as yet still to be applied at any appropriate scale include planting trees, dumping lime into the ocean, building chemical sponges or factories to pull gas out of the air, and using iron-fertilisation techniques in areas of the ocean that have a low level of iron. (Iron is required for plant growth.) It seems best to work with Mother Nature when it comes to fast and medium speeds for sequestering carbon out of the atmosphere, but the technology is just not there yet, and there are no easy solutions or guaranteed successes, which means there has been a lack of investment. We rely so heavily on the ocean and are already doing so much damage to it that it doesn't seem right to then turn to it (or to developing countries) to do our carbon-removal dirty work for us while we continue living by the standards that we have become used to.

There are, of course, some hyperactive salespeople involved in carbon sequestration. A big push in the late 1990s and 2000s towards biofuels (fuels produced by living organisms, such as bioethanol produced by plants) meant that food security in some countries was at risk, because the land that would normally have been used to grow

159

food was being prioritised for biofuels. Petroleum companies indulged in a lot of greenwashing, going on and on about the percentage of their fuel that came from biofuels. It's unfortunate that we don't have the level of agricultural education we need for people to realise that many countries around the world are actually facing food security issues in the next decade or two as the carbon is eroded from the soils. It is an unequal game for developing nations, and some governments may be risking the food security of their populations in order to make money from global legislation and subsidies for carbon cap and trade.

Alternative fuels such as hydrogen will require huge amounts of infrastructure. In many parts of the world, green hydrogen is now being touted as a carbon-neutral fuel. Green hydrogen is just hydrogen that has been produced by splitting water into its elements (oxygen and hydrogen) using only electricity generated from low-carbon sources, such as wind farms, solar energy or hydropower. This hydrogen can then be stored or transported as a new energy source (it can be simply burnt to create power, among other ways), negating the need for batteries.

And, hang on, what happened to electric vehicles (EVs)? We seem to have jumped straight to the idea of fuel cell vehicles (FCVs; electric vehicles with a fuel cell powered by hydrogen) rather than embracing what EVs can do. Some people believe that FCVs are actually the way we're going to go because EVs' need for batteries is going to make them unsustainable, and EVs are just a stepping stone to something better. Emissions from an FCV are actually just water and oxygen, but people see batteries as much less likely to explode than hydrogen tanks (remembering the *Hindenburg* disaster). We will need government incentives, bonuses and financial support in order to transfer away from petrol and diesel engines and actually use new and alternative types of vehicles.

Even a few years ago, in cities such as Amsterdam in the Netherlands, you could walk down the streets alongside the canals and see back-to-back Teslas all over the road. This was because the Dutch government (and others across the EU) provided incentive programs in which you would get money for returning your old car, and a bonus and a reduction in cost for accessing an EV. Interestingly, the trams were also electrified. I think the only combustion engine I directly witnessed was the tourist boat that chugged me around the canals for a tour.

In the next few years, our 'petrol stations' will start to evolve away from petrol. The next big buzzword when it comes to fuel is e-fuels (electrofuels). These are an emerging area of alternative fuels that could drop straight into the current fuel supply chains and are made by storing energy from renewables as chemical energy. It was only in 2021 that the world finally stopped using leaded petrol. The environmental and human costs of leaded petrol were well known and taken seriously, but it took decades to completely move away from it. The move to unleaded petrol was generally socially accepted because it was so easy. We need to replicate that ease in the transition from traditional internal combustion engines to vehicles that use alternative methods.

Shipping and food miles

Shipping is much more significant than the aviation industry in terms of carbon emissions. Carbon emissions from shipping are actually increasing annually towards a predicted level of 17 per cent of total carbon emissions by 2050, and yet we never really consider this to be significant. We don't interact with ships in the way we do with buses or cars or planes; we don't really see them unless they're involved in a crash or disaster, such as an oil tanker spill, or the incident in 2021 of *Ever Given* getting stuck in the Suez Canal. Innovations in the shipping industry are starting to come online just as questions are being asked about its impacts. In the last few years, we have seen an autonomous, carbon-neutral shipping container transport undergoing trials in Norway to see whether or not it could be used as an alternative to the regular shipping vessels on scale.

As I discussed in chapter 7, there have been conversations over the past few decades about food miles – how far the ingredients in your food have travelled to get to your plate. Globalisation of food supply has been a silently evolving thing, and it can be hard to find local produce in the supermarket now. That's another reason attendance at local farmers markets has increased, especially in affluent areas where the demographic has extra cash and a concern about the climate. No plastic bags to be seen. The key again, though, is scalability and mass acceptance – as well as pricing that is egalitarian, to allow all to access it.

Current Qantas CEO Alan Joyce has stated that his dream is to have a carbon-negative business as soon as is feasible, including new technology and engineering on aircraft. Qantas has even done landfill-free flights (pre-COVID-19) to test the attitudes of passengers and the flexibility of the supply chain. It's really sad that some of the aviation companies that were trying to push the envelope on these changes have been hit so badly by the COVID-19 pandemic that they may not have the money to invest where it will be needed.

Something you can do when you're flying is to ensure that your journey is carbon-neutral through carbon offsets. Carbon offsets have sometimes been criticised as a great way for a corporation to try to make it look like that business is doing a better job for the environment than they actually are, but I'm a big proponent of people actually trying to do the right thing, even if it's not perfect. The more of us who choose to donate points or money to carbon storage, carbon capture and renewable energy projects, the more those projects get investment; and when those projects get investment, they can scale up; and when they can scale up, they can demonstrate how they can be the way forward. Ideas on the fringes now in terms of a carbon-negative economy could become normalised over the next ten years or so.

When you look at it in terms of our landmass and the various ways in which we could export the outputs from solar, wind and green hydrogen, Australia could potentially be a battery pack for the whole of the Asia-Pacific Region. We could put cables under the sea to pump electricity up to Singapore, or Indonesia, or across the Pacific Islands, and this could be a game changer in terms of the reduction of carbon emissions from local energy generation. It always surprises and some-times saddens me that many people who live in island nations don't already have a full suite of renewable energy technologies available and are still so reliant upon diesel-fuelled generators.

Carbon negative is the new carbon neutral

As we recognise the need to decarbonise our economies, it seems we can't go fast enough to prevent planet-changing levels of carbon in our atmosphere.

The climate has changed. Recently, Australia was burning, Siberia was on fire, Canada was ablaze, and California has also taken some heat. The major issue with severe fires is that they then feed atmospheric carbon levels, leading to more extreme events. When the Amazon rainforest had severe fires, everyone worried about the levels of oxygen production necessary for us all to breathe, acutely unaware that up to 80 per cent of our oxygen comes from the oceans (until the oceans get so hot that they stop producing it, of course).

Reef grief; media miscommunication; scientists screaming into the dark. The roof is on fire, and we need someone to call the fire brigade! But, seriously, who are we going to call?

Get back in the driver's seat

The climate emergency is becoming such a pressing economic issue that we have to work collectively on seemingly small changes to create large effects. Things you can do include switching energy suppliers, moving your superannuation if your current provider invests in fossil fuels, and being mindful of what you are buying and where from. But remember: we are up against some large industries that still don't fully recognise or pay for the effects that their business models have on climate change. It's time to start holding these companies to account by creating a carbon market that is designed to cancel out climate change. Build it with a core purpose, and let it fight eco-anxiety at the same time.

My good friend Dr Gill Hicks was severely injured in the 7/7 bombings in London in 2005, and so, unsurprisingly, her daughter gets anxious whenever she sees terrorism events in the media. Gill gives her daughter one piece of advice that I am going to pass to you like a torch in the dark: always look for the people who are helping.

To dream the impossible dream

'To be, or not to be, that is the question:
Whether 'tis nobler in the mind to suffer
The slings and arrows of outrageous fortune,
Or to take arms against a sea of troubles
And by opposing end them. To die – to sleep,
No more; and by a sleep to say we end
The heart-ache and the thousand natural shocks
That flesh is heir to: 'tis a consummation
Devoutly to be wish'd. To die, to sleep;
To sleep, perchance to dream – ay, there's the rub:
For in that sleep of death what dreams may come,
When we have shuffled off this mortal coil,
Must give us pause...'
– William Shakespeare, *Hamlet*

As Kurt Cobain sang, 'All in all is all we are'. This statement succinctly recognises that what we don't know is where we haven't yet looked. Your current boundaries (intellectual, spiritual, geographical) were once unknown places, and there is something to be said about an adventurous heart and spirit – it is insatiable, never satisfied. The future is fluid, and we can only handle it a teaspoon at a time. This is why it is so hard to think about the future and predict with much accuracy what will happen. As we increasingly know more and can change so many things, the present can turn on a penny.

So, is the future broken? Are we racing towards a brick wall like the supposed trap we can never escape, as described by Malthus, or are we able to self-rescue from the darkest of our fates through the power of technology combined with human ingenuity?

Homo sapiens is a confusing species. We attempt altruism and end up with consumerism. We are moral people, but we choose politicians who do not necessarily make the most moral decisions for our grandchildren. At the time of writing, we are watching with bated breath as to whether our societies will choose the status quo from before the COVID-19 pandemic, or if it's possible to dump some baggage.

The answers and languages we should be prioritising are those that come from STEM, because they adapt to new information, they evolve with currency, and they are always checking whether they are significant or not. This is why I trust in STEM and the people who choose to study these fields.

That said, there is a difference between surviving and living. Living means we need to especially understand what it means to have a soul. This is where the creative arts lead and I follow. My maternal great, great, great, great, great-grandfather was Edward Hodges Baily. I am sure you've never heard of him, but you may know some of his works, such as the statue of Horatio Nelson on the top of the biggest column in Trafalgar Square, London. Here is where science, art, history and patriotism met, and what a challenge. Sculpture is the area of the art world that I am most drawn to, likely because of the science and engineering involved as an active part of creativity.

And nature is creative, too. I like looking up when walking through a city and checking on the plants that shouldn't be there – the trees and grasses that grow out of drains or cracks in the cement. I guess it is because I have an affinity for the rebellious, and Mother

Nature is certainly a rebel – an untameable power, and a power that could destroy all we have created if we don't start respecting her more, and fast.

After this decade and the one following, we will move on to Industry 6.0, in which technology and nature coexist for the protection of all, powered by quantum computing and technologies we haven't even heard of yet (outside of whisperings in labs and philosophical chats in the pub).

To get to this utopia, we must take on or continue a few fights on a few fronts: social licence based on STEM, digital literacy and equality; technology existing for all and the monopoly of technology belonging to no one entity (we have to protect competition in this space); celebrating scientific achievements the way we currently celebrate sporting ones; and the need for the media, movies, TV, social media and new immersive technologies and games to capture the hearts and minds of everyone and be representative of the whole of society.

AI could help us cure cancer. AI could also potentially be involved in propaganda and control. We know that drones can deliver vaccines. Drones can also fire guns and kill people. When we look at planet Earth and the ocean, cutting-edge technology and solutions are not equally available around the world. When we look at food security, when we look at satellite communications, when we look at flying to Mars, not everybody has the same fair go.

So many of the emerging technologies we need to drive a positive future are already here. What do we want to do with them?

We have to get better at opening up education opportunities for people like you to come and experience what these technologies actually look like – what they could mean, what they could be – and also to challenge the assumptions around the economics, governance and politics of some of these technologies. Challenge it all.

The future is already here. Strive for what you want.

*'Here's to the crazy ones. The misfits, the rebels,
the troublemakers, the round pegs in the square holes ...
the ones who see things differently – they're not fond of rules,
and they have no respect for the status quo ... You can quote
them, disagree with them, glorify or vilify them, but the only
thing you can't do is ignore them because they change things ...
They push the human race forward, and while some
may see them as the crazy ones, we see genius, because
the people who are crazy enough to think that they can
change the world, are the ones who do.'*
– 'To the crazy ones', part of Apple's 'Think Different'
advertising campaign

About the author

Dr Catherine Ball is an academic, an entrepreneur, a mother and a bit of a rebel. A global visioneer for XPRIZE on climate and environment, an advisor to start-ups, scale-ups and organisations changing the world for the better, Dr Cath is a thought leader who is constantly learning and evolving across global projects. An internationally award-winning businesswoman, an active ally and a loyal friend, Dr Cath can normally be found reading books, having passionate discussions, supporting the arts, backing womens' sports and hanging out with her young family. You can connect with her at www.DrCatherineBall.com.

Acknowledgements

A special welcome to anyone who reads the acknowledgements before reading the book itself – hello there, you rebel, I like the cut of your jib.

Big thanks to my editorial and publishing teams, without whom these words would not be here.

I couldn't do anything interesting or demanding of my time without the love and support of my husband, Jeremy, aka J-Dawg. Thank you for being my biggest supporter and the best dad I could have ever dreamed of for my sons. I am so glad I commingled my DNA with yours.

To my boys, Nate and Nick, who make me laugh in ways I never knew I could. I love you both to infinity and beyond. You've been the saving of me, and the making of me as a mother.

To my family, the ones who raised me and loved me unconditionally. I hope I have made you proud – especially you, Mum.

To my mates oldest and newest, and the men and women who open doors for me and lift me up at every opportunity, you're the brothers and sisters I never had – especially you, Brit.

To Jyotika and Genevieve, who have shown me so many new horizons and shone a spotlight on me so that others would notice and take interest – I really will never be able to repay you for your kindness and acknowledgements. Thank you both.

To my future family, far and wide in time and space, and all that the future connection of the next generations could be.

And to you, dear reader – see you in the metaverse!

References

Introduction

R.L. Martin & A. Kemper, 'Saving the planet: a tale of two strategies', *Harvard Business Review*, April 2012, <hbr.org/2012/04/saving-the-planet-a-tale-of-two-strategies>.

C. Sommers, *Think like a futurist: know what changes, what doesn't, and what's next*, Jossey-Bass, 2012.

Chapter 1

O. Wilde, 'The Remarkable Rocket', *The Happy Prince and Other Tales*, David Nutt, London, 1888.

R. Descartes, *A discourse on method*, J.M. Dent, London, 1920.

J. Brown, 'Why "phubbing" is killing your relationship', *Fatherly*, 11 September 2018, <fatherly.com/love-money/phubbing-ruining-relationships-phone-snubbing>.

M. van den Heuvel, J. Ma, C.M. Borkhoff, C. Koroshegyi, D.W.H. Dai, P.C. Parkin, J.L. Maguire & C.S. Birken on behalf of the TARGet Kids! Collaboration, 'Mobile media device use is associated with expressive language delay in 18-month-old children', *Journal of Developmental & Behavioral Pediatrics*, 2019, vol. 40, no. 2, pp. 99–104.

G. Citroner, 'Excessive screen time for kids can cause developmental delays by kindergarten', *healthline*, 27 January 2019, <healthline.com/health-news/does-screen-time-cause-developmental-delays-in-young-children>.

M. Gregory, 'What does the "still face" experiment teach us about connection?', PsychHelp, <psychhelp.com.au/what-does-the-still-face-experiment-teach-us-about-connection/>, accessed 21 July 2022.

N. Carr, *The shallows: what the internet is doing to our brains*, W.W. Norton, New York, 2010.

D.R. Papke, 'Karl Marx on religion', *Marquette University Law School Faculty Blog*, 20 January 2015, <law.marquette.edu/facultyblog/2015/01/karl-marx-on-religion/comment-page-1/>.

M. Barlow & C. Lévy-Bencheton, 'The smart nation where everyone owns their personal data', SmartCitiesWorld, 24 October 2018, <smartcitiesworld.net/special-reports/special-reports/the-smart-nation-where-everyone-owns-their-personal-data>.

C. Knaus, 'More than 2.5 million people have opted out of My Health Record', *The Guardian*, 20 February 2019, <theguardian.com/australia-news/2019/feb/20/more-than-25-million-people-have-opted-out-of-my-health-record>.

C. Ozdoruk, 'Do people prefer chatbots or humans? It depends', *readwrite*, 5 March 2021, <readwrite.com/do-people-prefer-chatbots-or-humans-it-depends/>.

Minority Report, motion picture, 2002.

D. Mitchell, *Cloud Atlas*, Sceptre, London, 2004.

W.D. Heaven, 'Predictive policing algorithms are racist. They need to be dismantled', MIT *Technology Review*, 17 July 2020, <technologyreview.com/2020/07/17/1005396/predictive-policing-algorithms-racist-dismantled-machine-learning-bias-criminal-justice/>.

J. Worthington, 'Tasmania missing from unused piece of COVIDSafe source code', *The Examiner*, 11 May 2020, <examiner.com.au/story/6752933/tasmania-missing-from-covidsafe-code/>.

Chapter 2

'Society 5.0', Cabinet Office, Government of Japan, <www8.cao.go.jp/cstp/english/society5_0/index.html>, accessed 21 July 2022.

'Revealed: plastic ingestion by people could be equating to a credit card a week', WWF Australia, 12 June 2019, <wwf.org.au/news/news/2019/revealed-plastic-ingestion-by-people-could-be-equating-to-a-credit-card-a-week>.

'Does your board have the STEM skills it needs?' Australian Institute of Company Directors, 18 May 2020, <aicd.com.au/board-of-directors/performance/skills-matrix/does-your-board-have-the-stem-skills-it-needs.html>.

TEDx Talks, 'Designing cities for women | Lucy Turnbull | TEDxMacquarieUniversity', video, YouTube, 26 November 2014, <youtube.com/watch?v=oUFxDXB24J0>.

S. Lane, 'The scary facts behind the gender pension gap', World Economic Forum, 7 March 2018, <weforum.org/agenda/2018/03/retired-women-less-money-pensions-than-men/>.

D. Warren, 'Families in Australia Survey report', Australian Institute of Family Studies, September 2021, <aifs.gov.au/research/research-reports/towards-covid-normal-early-release-superannuation-through-family-lens>.

J. Joly, 'Four-day week: which countries have embraced it and how's it going so far?', euronews.next, updated 6 June 2022, <euronews.com/next/2022/06/06/the-four-day-week-which-countries-have-embraced-it-and-how-s-it-going-so-far>.

E. Ainge Roy, 'Jacinda Ardern flags four-day working week as way to rebuild New Zealand after Covid-19', *The Guardian*, 20 May 2020, <theguardian.com/world/2020/may/20/jacinda-ardern-flags-four-day-working-week-as-way-to-rebuild-new-zealand-after-covid-19>.

R. Sutherland, 'The TWaT revolution: office on Tuesday, Wednesday and Thursday only', *The Spectator*, 19 January 2019, <spectator.co.uk/article/the-twat-revolution-office-on-tuesday-wednesday-and-thursday-only>.

R. de Wit & T. Bouvier, '"Everything is everywhere, but, the environment selects"; what did Baas Becking and Beijernick really say?', *Environmental Microbiology*, 2006, vol. 8, no. 4, pp. 775-758.

H. Dediu, 'Steve Jobs's ultimate lesson for companies', *Harvard Business Review*, 25 August 2011, <hbr.org/2011/08/steve-jobss-ultimate-lesson-fo>.

A. Rose, '*Forgery* – Australasian Dance Collective dance against the machine', *scenestr*, 19 August 2021, <scenestr.com.au/arts/forgery-australasian-dance-collective-dance-against-the-machine-20210819>.

G. Bell, '6 big ethical questions about the future of AI', video, TED Salon: Dell Technologies, <ted.com/talks/genevieve_bell_6_big_ethical_questions_about_the_future_of_ai>, accessed 21 July 2022.

G. Bell, 'Fast, smart and connected: can we be Australian in a digital world?', Boyer Lectures, ABC Ultimo, Sydney, 21 October 2017.

Chapter 3

Blade Runner, motion picture, 1982.

Audacy Originals, 'Rotary phone stumps two teenage boys in hilarious bet', video, YouTube, 12 January 2019, <youtube.com/watch?v=n0geurEskdo>.

C. Summers, 'Mobile phones – the new fingerprints', *BBC News Online*, updated 18 December 2003, <news.bbc.co.uk/2/hi/uk_news/3303637.stm>.

R. Armitage, 'The Golden State Killer: how it took four decades to catch the serial predator who terrorised California', *ABC News*, 22 August 2020, <abc.net.au/news/2020-08-22/the-forty-year-hunt-for-the-golden-state-killer/12579638>.

S. Knapton, 'Criminals could alter their DNA to evade justice with new genetic editing tools', *The Telegraph*, 5 May 2018, <telegraph.co.uk/science/2018/05/05/criminals-could-alter-dna-evade-justice-new-genetic-editing/>.

K. Hill, 'How Target figured out a teen girl was pregnant before her father did', *Forbes*, 16 February 2012, <forbes.com/sites/kashmirhill/2012/02/16/how-target-figured-out-a-teen-girl-was-pregnant-before-her-father-did/?sh=190f79676668>.

D. Hambling, 'The Pentagon has a laser that can identify people from a distance – by their heartbeat', *MIT Technology Review*, 27 June 2019, <technologyreview.com/2019/06/27/238884/the-pentagon-has-a-laser-that-can-identify-people-from-a-distanceby-their-heartbeat/>.

N. Confessore, 'Cambridge Analytics and Facebook: the scandal and the fallout so far', *The New York Times*, 4 April 2018, <nytimes.com/2018/04/04/us/politics/cambridge-analytica-scandal-fallout.html>.

N. Jones, 'How to stop data centres from gobbling up the world's electricity', *nature*, 12 September 2018, <nature.com/articles/d41586-018-06610-y>.

L. Harding, 'The node pole: inside Facebook's Swedish hub near the Arctic Circle', *The Guardian*, 25 September 2015, <theguardian.com/technology/2015/sep/25/facebook-datacentre-lulea-sweden-node-pole>.

T. Dawn-Hiscox, 'Facebook plans third data center in Luleå, Sweden' *Data Center Dynamics*, 8 May 2018, <datacenterdynamics.com/en/news/facebook-plans-third-data-center-in-lule%C3%A5-sweden/>.

'NEXTDC', Climate Active, <climateactive.org.au/buy-climate-active/certified-members/nextdc>, accessed 21 July 2022.

L. Bradshaw, 'Big data and what it means', U.S. Chamber of Commerce Foundation, <uschamberfoundation.org/bhq/big-data-and-what-it-means>, accessed 21 July 2022.

'ACSC annual cyber threat report: July 2019 to June 2020', Australian Cyber Security Centre, <cyber.gov.au/sites/default/files/2020-09/ACSC-Annual-Cyber-Threat-Report-2019-20.pdf>, accessed 21 July 2022.

C. Faife, 'The Marriott hotel chain has been hit by another data breach', *The Verge*, 6 July 2022, <theverge.com/2022/7/6/23196805/marriott-hotels-maryland-data-breach-credit-cards>.

J. Hendry, 'Oxfam Australia creates CDO role after data breach', *iTnews*, 20 August 2021, <itnews.com.au/news/oxfam-australia-creates-cdo-role-after-data-breach-568601>.

M. Baker, 'Everything you need to know about the MyFitnessPal data breach', *UK Tech News*, 1 October 2021, <uktechnews.co.uk/2021/10/01/everything-you-need-to-know-about-the-myfitnesspal-data-breach/>.

R. Crozier, 'Canva's infosec resourcing "still growing" two years after large data breach', *iTnews*, 2 September 2021, <itnews.com.au/news/canvas-infosec-resourcing-still-growing-two-years-after-large-data-breach-569282>.

';--have i been pwned?, <haveibeenpwned.com>, accessed 21 July 2022.

J. Weaver & S. O'Connor, *Tending the tech-ecosystem: who should be the tech-regulator(s)?*, Tech Policy Design Centre, Australian National University, May 2022.

ySafe, <ysafe.com.au>, accessed 15 August 2022.

'Digital Technologies', <australiancurriculum.edu.au/f-10-curriculum/technologies/digital-technologies/?>, accessed 15 August 2022.

Chapter 4

'Three laws of robotics', Google Arts & Culture, <artsandculture.google.com/entity/three-laws-of-robotics/m0gcd2>, accessed 21 July 2022.

I. Asimov, *Foundation and Earth*, Doubleday, Garden City (N.Y.), 1986.

'Buy less, choose well, make it last', B.e Quality, 20 May 2020, <be-quality.com/en/buy-less-choose-well-make-it-last-vivienne-westwood/>.

J. Allchin, 'Case study: Patagonia's "Don't buy this jacket" campaign', *MarketingWeek*, 23 January 2013, <marketingweek.com/case-study-patagonias-dont-buy-this-jacket-campaign/>.

'A new textiles economy: redesigning fashion's future', Ellen MacArthur Foundation, <ellenmacarthurfoundation.org/a-new-textiles-economy>, accessed 21 July 2022.

B. Balcer, 'Exoskeletons are giving Japanese senior workers a lift', Pop-Up City, 26 May 2020, <popupcity.net/observations/exoskeletons-are-giving-japanese-senior-workers-a-lift/>.

References

'Cost of injury and illness by type', Safe Work Australia, <safeworkaustralia.gov.au/data-and-research/work-related-injuries/cost-injury-and-illness-type>, accessed 21 July 2022.

B. Debusmann Jr., 'The "Iron Man" body armour many of us may soon be wearing', *BBC News*, 12 April 2021, <bbc.com/news/business-56660644>.

F.H.M. van Herpen, R.B. van Dijsseldonk, H. Rijken, N.L.W. Keijsers, J.W.K. Louwerens & I.J.W. van Nes, 'Case report: description of two fractures during the use of a powered exoskeleton', *Spinal Cord Series and Cases*, 2019, vol. 5, no. 99, published online, <nature.com/articles/s41394-019-0244-2>.

How to manage work health and safety risks: code of practice, Safe Work Australia, December 2011, <safeworkaustralia.gov.au/system/files/documents/1702/how_to_manage_whs_risks.pdf>.

Chapter 5

'IOC EB recommends no participation of Russian and Belarusian athletes and officials', International Olympic Committee, 28 February 2022, <olympics.com/ioc/news/ioc-eb-recommends-no-participation-of-russian-and-belarusian-athletes-and-officials>.

A View to a Kill, motion picture, 1985.

Innerspace, motion picture, 1987.

CYBATHLON, <cybathlon.ethz.ch>, accessed 21 July 2022.

L. Reinhardt, R. Scheswig, A. Lauenroth, S. Schulze & E. Kurz, 'Enhanced sprint performance analysis in soccer: new insights from a GPS-based tracking system', PLoS ONE, 2019, vol. 14, no. 5, published online, <journals.plos.org/plosone/article?id=10.1371/journal.pone.0217782>.

S. Farrell, 'Former USWNT players lead push to study CTE and headers', Global Sport Matters, 30 July 2019, <globalsportmatters.com/health/2019/07/30/cte-study-former-uswnt-players-headers/>.

D.B. Taylor & N. Chokshi, 'This Fortnite World Cup winner is 16 and $3 million richer', *The New York Times*, 29 July 2019, <nytimes.com/2019/07/29/us/fortnite-world-cup-winner-bugha.html>.

S. Dredge, 'Facebook closes its $2bn Oculus Rift acquisition. What next?', *The Guardian*, 22 July 2014, <theguardian.com/technology/2014/jul/22/facebook-oculus-rift-acquisition-virtual-reality>.

N. James, 'Secrets of how F1 drivers prepare for brand new tracks like Austrian Grand Prix', *Bleacher Report*, 17 June 2014, <bleacherreport.com/articles/2099152-secrets-of-how-f1-drivers-prepare-for-brand-new-tracks-like-austrian-grand-prix>.

Robot Wars, television program, 1998–2004 & 2016–2018.

B. Marr, 'A short history of the metaverse', *Forbes*, 21 March 2022, <forbes.com/sites/bernardmarr/2022/03/21/a-short-history-of-the-metaverse/?sh=2fe8e9df5968>.

Inception, motion picture, 2010.

RoboCup, <robocup.org>, accessed 21 July 2022.

C. Krauthammer, 'Be afraid', *The Weekly Standard*, 26 May 1997, <washingtonexaminer.com/weekly-standard/be-afraid-9802>.

S. Byford, 'AlphaGo retires from competitive Go after defeating world number one 3–0', *The Verge*, 27 May 2017, <theverge.com/2017/5/27/15704088/alphago-ke-jie-game-3-result-retires-future>.

WarGames, motion picture, 1983.

Chapter 6

W. Blake, *The Marriage of Heaven and Hell*, Camden Hotten, London, 1868.

T.D. Crouch, 'Samuel Pierpoint Langley: American engineer', *Britannica*, 26 July 1999, <britannica.com/biography/Samuel-Pierpont-Langley>.

M. Beschloss, 'Marilyn Monroe's World War II Drone Program', *The New York Times*, 3 June 2014, <nytimes.com/2014/06/04/upshot/marilyn-monroes-world-war-ii-drone-program.html>.

'GAF Jindivik', *Wikipedia*, <en.wikipedia.org/wiki/GAF_Jindivik>, accessed 21 July 2022.

'Zephyr: the first stratospheric UAS of its kind', Airbus, <airbus.com/defence/uav/zephyr.html>, accessed 21 July 2022.

L. Hutchinson, 'How NASA steers the International Space Station around space junk', *Ars Technica*, 5 July 2013, <arstechnica.com/science/2013/07/how-nasa-steers-the-international-space-station-around-space-junk/>.

Economic benefit analysis of drones in Australia, Deloitte, 23 October 2020, <www2.deloitte.com/au/en/pages/public-sector/articles/economic-benefit-analysis-drones-australia.html>.

A. Moses, 'Here comes the Drone Age', *The Sydney Morning Herald*, 11 September 2012, <smh.com.au/technology/here-comes-the-drone-age-20120910-25o6p.html>.

'DS30W', Doosan Mobility innovation, <doosanmobility.com/en/products/drone-ds30/>, accessed 21 July 2022.

'In a race to save lives post-cardiac arrest, drones may beat ambulances 93% of the time', Advisory Board, 19 June 2017, <advisory.com/en/daily-briefing/2017/06/19/aed-drones>.

Gravity Industries, <gravity.co>, accessed 21 July 2022.

J. Gorzelany, 'Volvo will accept liability for its self-driving cars', *Forbes*, 9 October 2015, <forbes.com/sites/jimgorzelany/2015/10/09/volvo-will-accept-liability-for-its-self-driving-cars/?sh=4627d80d72c5>.

'Toyota investing $400 million in flying car company', *Tech Xplore*, 16 January 2020, <techxplore.com/news/2020-01-toyota-investing-million-car-company.html>.

'Audi, Airbus and Italdesign test flying taxi concept', video, Audi MediaTV, 27 November 2018, <audi-mediacenter.com/en/audimediatv/video/audi-airbus-and-italdesign-test-flying-taxi-concept-4425>.

J. Peskett, 'Singapore trials first UAV traffic management system', *Commercial Drone Professional*, 29 March 2021, <commercialdroneprofessional.com/singapore-trials-first-uav-traffic-management-system/>.

References

Skyeton, 'Long-range drone/UAV for anti-poaching and wildlife conservation', Geo-matching, <geo-matching.com/content/long-range-drone/uav-for-anti-poaching-and-wildlife-conservation>, accessed 21 July 2022.

'AI and cloud combine to help protect vulnerable marine turtle populations in northern Australia and Cape York', Microsoft News Center, 18 February 2021, <news.microsoft.com/en-au/features/turtles-and-ai/>.

L. Lippsett, 'Gliders tracked potential for oil to reach the east coast', *Oceanus*, 23 March 2011, <whoi.edu/oceanus/feature/gliders-tracked-potential-for-oil-to-reach-the-east-coast/>.

Saildrone, <saildrone.com>, accessed 21 July 2022.

D. Bressan, 'Coastal erosion is accelerating at alarming rates and humans are to blame', *Forbes*, 12 November 2016, <forbes.com/sites/davidbressan/2016/11/12/coastal-erosion-is-accelerating-at-worrying-rate-and-human-activity-is-to-blame/?sh=5fcb2f3756cd>.

'FAIR principles', GO FAIR, <go-fair.org/fair-principles/>, accessed 21 July 2022.

The SkyBound Rescuer Project, <skyboundrescuerproject.com>, accessed 21 July 2022.

T. Hornyak, 'Secom security drone follows, photographs intruders', CSO, 22 May 2015, <csoonline.com/article/2925641/secom-security-drone-follows-photographs-intruders.html>.

R. Nordland, 'Grenfell Tower: firefighters say ladders came up short and cost lives', *The Sydney Morning Herald*, 9 July 2017, <smh.com.au/world/grenfell-tower-firefighters-say-ladders-came-up-short-and-cost-lives-20170709-gx7ixo.html>.

M. Margaritoff, 'A drone helped firefighters combat the London Grenfell Tower inferno', *The Drive*, 20 June 2017, <thedrive.com/article/11701/a-drone-helped-firefighters-combat-the-london-grenfell-tower-inferno>.

World of Drones & Robotics Global, <worldofdronesandrobotics.com>, accessed 21 July 2022.

A. Thorn, 'World-first as Google drones deliver from Queensland mall', *Australian Aviation*, 7 October 2021, <australianaviation.com.au/2021/10/world-first-as-google-drones-deliver-from-queensland-mall/>.

A. Stewart & D. Friesen, 'Special drone delivery: the world-renowned Canadian taking organ transplants to new heights', *Global News*, 5 February 2022, <globalnews.ca/news/8583562/drone-delivery-organ-transplants-canadian/>.

E. Snouffer, 'Six places where drones are delivering medicines', *nature*, 13 April 2022, <nature.com/articles/d41591-022-00053-9>.

'Drone racing', *Wikipedia*, <en.wikipedia.org/wiki/Drone_racing>, accessed 21 July 2022.

'Toddler's eyeball sliced in half by drone propeller', *BBC News*, 26 November 2015, <bbc.com/news/uk-england-hereford-worcester-34936739>.

E. Zhan, '3,281 drones break dazzling record for most airborne simultaneously', *Guinness World Records*, 17 May 2021, <guinnessworldrecords.com/news/commercial/2021/5/3281-drones-break-dazzling-record-for-most-airborne-simultaneously-655062>.

A. Burke, 'After Pirates of the Caribbean, drones are changing how blockbusters are made', *Financial Review*, 27 May 2017, <afr.com/life-and-luxury/arts-and-culture/drones-reshaping-the-way-blockbusters-are-made-20170526-gwe9fg>.

'XM2 TANGO Accessories – Heavy Lift UAV', XM2, <xm2store.com.au/collections/tango-drone-accessories>, accessed 21 July 2022.

'Cat drone inventor works on flying cows', *BBC News*, 3 August 2016, <bbc.com/news/technology-36954689>.

R. Brennan, 'The bizarre items banned for G20', *The Courier Mail*, 12 November 2014, <couriermail.com.au/news/queensland/the-bizarre-items-banned-for-g20/news-story/2d44b174c8a230a3926edb7e33c33649>.

'Drones caught in no-fly zones on the Gold Coast', *Triple M*, 9 April 2018, <triplem.com.au/story/drones-caught-in-no-fly-zones-on-the-gold-coast-89139>.

Y. Redrup, 'DroneShield guns beef up Commonwealth Games security arsenal', *Financial Review*, 5 April 2018, <afr.com/technology/droneshield-guns-beef-up-commonwealth-games-security-arsenal-20180405-h0yda3>.

T. Ong, 'Dutch police will stop using drone-hunting eagles since they weren't doing what they were told', *The Verge*, 12 December 2017, <theverge.com/2017/12/12/16767000/police-netherlands-eagles-rogue-drones>.

'Drone "containing radiation" lands on roof of Japanese PM's office', *The Guardian*, 22 April 2015, <theguardian.com/world/2015/apr/22/drone-with-radiation-sign-lands-on-roof-of-japanese-prime-ministers-office>.

Chapter 7

B. Potter, 'The Tale of Peter Rabbit', *Peter Rabbit and Other Stories*, 1902, Lit2Go Edition, <etc.usf.edu/lit2go/148/peter-rabbit-and-other-stories/4923/the-tale-of-peter-rabbit/>, accessed 21 July 2022.

'Growing at a slower pace, world population is expected to reach 9.7 billion in 2050 and could peak at nearly 11 billion around 2100', United Nations Department of Economic and Social Affairs, 17 June 2019, <un.org/development/desa/en/news/population/world-population-prospects-2019.html>.

'Worldwide food waste', ThinkEatSave, UN Environment Programme, <unep.org/thinkeatsave/get-informed/worldwide-food-waste>, accessed 21 July 2022.

LiveAid1, 'BBC News 10/23/84' ⊕ Michael Buerk (Highest Quality)', video, YouTube, 14 November 2009, <youtube.com/watch?v=XYOj_6OYuJc>.

A. de Waal, *Evil days: thirty years of war and famine in Ethiopia*, Human Rights Watch, New York & London, 1991.

'Yemen emergency', UN World Food Programme, <wfp.org/emergencies/yemen-emergency>, accessed 21 July 2022.

C. Arsenault, 'Only 60 years of farming left if soil degradation continues', *Scientific American*, 5 December 2014, <scientificamerican.com/article/only-60-years-of-farming-left-if-soil-degradation-continues/>.

References

K.M. Bailey, 'Monterey Bay abalone farm shows what sustainable aquaculture can be like', *Earth Island Journal*, 12 March 2015, <earthisland.org/journal/index.php/articles/entry/monterey_bay_abalone_farm_shows_what_sustainable_aquaculture_can_be_like/>.

A. Shapiro, 'The perfect frozen veggie burger doesn't exi—', *Bon Appétit*, 4 June 2021, <bonappetit.com/story/akua-kelp-burger>.

'Food miles', *Wikipedia*, <en.wikipedia.org/wiki/Food_miles>, accessed 21 July 2022.

H.L.I. Bornett, J.H. Guy & P.J. Cain, 'Impact of animal welfare on costs and viability of pig production in the UK', *Journal of Agricultural and Environmental Ethics*, 2003, vol. 16, no. 2, pp. 163–186.

C. Marshall & M. Prior, 'Call for higher animal welfare standards for Parma ham pigs', *BBC News*, 7 July 2022, <bbc.com/news/science-environment-62065102>.

'The common agricultural policy at a glance', Agriculture and rural development, European Commission, <ec.europa.eu/info/food-farming-fisheries/key-policies/common-agricultural-policy/cap-glance_en>, accessed 21 July 2022.

'2001 United Kingdom foot-and-mouth outbreak', *Wikipedia*, <en.wikipedia.org/wiki/2001_United_Kingdom_foot-and-mouth_outbreak>, accessed 21 July 2022.

N. Watson, J-P. Brandel, A. Green, P. Hermann, A. Ladogana, T. Lindsay, J. Mackenzie, M. Pocchiari, C. Smith, I. Zerr & S. Pal, 'The importance of ongoing international surveillance for Creutzfeldt-Jakob disease', *Nature Reviews Neurology*, 2021, vol. 17, pp. 362–379, <nature.com/articles/s41582-021-00488-7>.

'Bovine spongiform encephalopathy', *Wikipedia*, <en.wikipedia.org/wiki/Bovine_spongiform_encephalopathy>, accessed 21 July 2022.

J. Loria, 'There are more than 1.5 billion cows on the planet. Here's why that's a problem', blog, Mercy for Animals, 4 August 2017, <mercyforanimals.org/blog/there-are-more-than-15-billion-cows-on-the/>.

M. Conrad Stöppler (ed.), 'Medical definition of Frankenfood', *MedicineNet*, reviewed 29 March 2021, <medicinenet.com/frankenfood/definition.htm>.

F. Kools, 'What's been going on with the "hamburger professor"', Maastricht University, 11 April 2019, <maastrichtuniversity.nl/news/what's-been-going-'hamburger-professor'>.

The Chicken, <thechicken.kitchen/>, accessed 21 July 2022.

'Singapore restaurant is first to add lab-grown chicken to the menu', *Restaurant Technology News*, 2 January 2021, <restauranttechnologynews.com/2021/01/singapore-restaurant-is-first-to-add-lab-grown-chicken-to-the-menu/>.

Meat Free Monday <meatfreemondays.com>, accessed 21 July 2022.

'Meat free Monday', Mary McCartney, <marymccartney.com/food/meat-free-monday/>, accessed 21 July 2022.

G. Hill, 'Why I'm a weekday vegetarian', video, TED2010, <ted.com/talks/graham_hill_why_i_m_a_weekday_vegetarian>, accessed 21 July 2022.

Monster Kitchen, <monsterkitchen.com.au>, accessed 21 July 2022.

Veganuary, <veganuary.com>, accessed 21 July 2022.

D. Boffey, 'Robotic bees could pollinate plants in case of insect apocalypse', *The Guardian*, 9 October 2018, <theguardian.com/environment/2018/oct/09/robotic-bees-could-pollinate-plants-in-case-of-insect-apocalypse>.

'NSW issues bee lockdown after deadly varroa mite parasite discovered', *The Guardian*, 27 June 2022, <theguardian.com/environment/2022/jun/27/nsw-issues-bee-lockdown-after-deadly-parasite-discovered>.

T. Heyden, 'The cows that queue up to milk themselves', *BBC News*, 7 May 2015, <bbc.com/news/magazine-32610257>.

The Yield, <theyield.com>, accessed 21 July 2022.

N. Welti & R. McAllister, 'Verifying food credentials', CSIRO, updated 12 May 2022, <csiro.au/en/about/challenges-missions/trusted-agrifood-exports/building-an-australian-food-provenance-infrastructure>.

P. Wallace, 'Toowoomba export hub at Wellcamp Airport takes off', *Daily Cargo News*, 27 July 2021, <thedcn.com.au/news/bulk-trades-shipping/toowoomba-export-hub-at-wellcamp-airport-takes-off/>.

'Protection of the Champagne name', Comité Champagne, <champagne.fr/en/comite-champagne/bureaus/bureaus/united-states/pages/protection-of-the-champagne-name>, accessed 21 July 2022.

'Victory garden', *Wikipedia*, <en.wikipedia.org/wiki/Victory_garden>, accessed 21 July 2022.

S. Kaplan, 'A third of all food in the U.S. gets wasted. Fixing that could help fight climate change.', *The Washington Post*, 25 February 2021, <washingtonpost.com/climate-solutions/2021/02/25/climate-curious-food-waste/>.

'Demand growing – but you can help', Foodbank, <foodbank.org.au/growing-demand/>, accessed 21 July 2022.

Chapter 8

Gattaca, motion picture, 1997.

L. Poon, 'How cities are using digital twins like a SimCity for policymakers', *Bloomberg*, 6 April 2022, <bloomberg.com/news/features/2022-04-05/digital-twins-mark-cities-first-foray-into-the-metaverse>.

'NIH launches clinical trial of three mRNA HIV vaccines', National Institutes of Health, 14 March 2022, <nih.gov/news-events/news-releases/nih-launches-clinical-trial-three-mrna-hiv-vaccines>.

D. Wallace-Wells, 'We had the vaccine the whole time', *New York Magazine*, 7 December 2020, <nymag.com/intelligencer/2020/12/moderna-covid-19-vaccine-design.html>.

O. Schwartz, 'The rise of microchipping: are we ready for technology to get under our skin?', *The Guardian*, 8 November 2019, <theguardian.com/technology/2019/nov/08/the-rise-of-microchipping-are-we-ready-for-technology-to-get-under-the-skin>.

Total Recall, motion picture, 2012.

D. Marshall, 'Paralysis treatment closer after donation funds Australian-first research trial', *Griffith News*, 6 November 2020, <news.griffith.edu.au/2020/11/06/paralysis-treatment-closer-after-donation-funds-australian-first-research-trial/>.

References

A.T. Popescu, O. Stan & L. Miclea, '3D printing bone models extracted from medical imaging data', *2014 IEEE International Conference on Automation, Quality and Testing, Robotics*, 2014, pp. 1–5 <researchgate.net/publication/271484332_3D_printing_bone_models_extracted_from_medical_imaging_data>.

'Professor Mia Woodruff', Queensland University of Technology, <qut.edu.au/about/our-people/academic-profiles/mia.woodruff>, accessed 21 July 2022.

Richard P. Feynman, 'Plenty of room at the bottom', lecture to the American Physical Society in Pasadena, December 1959, <web.pa.msu.edu/people/yang/RFeynman_plentySpace.pdf>.

'Global nanomedicine market to be worth $258 billion by 2025, says report', *The Pharma Letter*, 27 May 2021, <thepharmaletter.com/article/global-nanomedicine-market-to-be-worth-258-billion-by-2025-says-report>.

The Andromeda Strain, motion picture, 1971.

'The search for new antibiotics', The Pew Charitable Trusts, 28 July 2017, <pewtrusts.org/en/research-and-analysis/articles/2017/07/the-search-for-new-antibiotics>.

'Antibiotic "last line of defence" breached in China', *CBC News*, 19 November 2015, <cbc.ca/news/health/antibiotic-resistance-colistin-1.3325942>.

'mRNA vaccines', Centers for Disease Control and Prevention, updated 15 July 2022, <cdc.gov/coronavirus/2019-ncov/vaccines/different-vaccines/mrna.html>.

D. Garde, 'The story of mRNA: how a once-dismissed idea became a leading technology in the Covid vaccine race', *STAT*, 10 November 2020, <statnews.com/2020/11/10/the-story-of-mrna-how-a-once-dismissed-idea-became-a-leading-technology-in-the-covid-vaccine-race/>.

J. Khan, 'We've never made a successful vaccine for a coronavirus before. This is why it's so difficult', *ABC News*, 17 April 2020, <abc.net.au/news/health/2020-04-17/coronavirus-vaccine-ian-frazer/12146616>.

'Human genome project timeline of events', National Human Genome Research Institute, updated 5 July 2022, <genome.gov/human-genome-project/Timeline-of-Events>.

myDNA, <mydna.life>, accessed 21 July 2022.

'NHS launches world first trial for new cancer test', NHS England, 13 September 2021, <england.nhs.uk/2021/09/nhs-launches-world-first-trial-for-new-cancer-test/>.

'Early detection of ovarian cancer', Gynaecological Cancer Research Group, <gyncancerresearch.org/early-detection>, accessed 21 July 2022.

Microba, <insight.microba.com>, accessed 21 July 2022.

Chapter 9

R. McKie, 'David Attenborough: force of nature', *The Guardian*, 28 October 2012, <theguardian.com/tv-and-radio/2012/oct/26/richard-attenborough-climate-global-arctic-environment>.

M. Crichton, *Jurassic Park*, Ballantine Books, New York, 1991.

'Extinction of thylacine', National Museum of Australia, <nma.gov.au/defining-moments/resources/extinction-of-thylacine>, accessed 21 July 2022.

'The mammoth', Colossal Laboratories & Biosciences, <colossal.com/mammoth/>, accessed 21 July 2022.

University of New South Wales, 'Scientists produce cloned embryos of extinct frog', *ScienceDaily*, 15 March 2013, <sciencedaily.com/releases/2013/03/130315151044.htm>.

M.L. Paúl, 'A woman cloned her pet after it died. But it's not a copycat.' *The Washington Post*, 11 April 2022, <washingtonpost.com/nation/2022/04/11/clone-pets-cat-science/>.

'Assisted evolution', Australian Institute of Marine Science, <aims.gov.au/reef-recovery/assisted-evolution>, accessed 21 July 2022.

R. Fearon, 'New reef discovery in Australia is a once in a century find', *Discovery*, 14 December 2020, <discovery.com/nature/coral-mountain--the-monumental-reef-discovered-on-the-great-barr>.

'Christina Kellogg, Ph.D.', United States Geological Survey, <usgs.gov/staff-profiles/christina-kellogg>, accessed 21 July 2022.

S. Brooke, 'Deepwater corals thrive at the bottom of the ocean, but can't escape human impacts', *The Conversation*, 3 December 2018, <theconversation.com/deepwater-corals-thrive-at-the-bottom-of-the-ocean-but-cant-escape-human-impacts-104211>.

The Nippon Foundation-GEBCO Seabed 2030 Project, <seabed2030.org>, accessed 21 July 2022.

'Environmental DNA', *Wikipedia*, <en.wikipedia.org/wiki/Environmental_DNA>, accessed 21 July 2022.

'Water Framework Directive', *Wikipedia*, <en.wikipedia.org/wiki/Water_Framework_Directive>, accessed 21 July 2022.

J. Li (ed.), 'Aims and scope', *International Journal of Applied Earth Observation and Geoinformation*, <journals.elsevier.com/international-journal-of-applied-earth-observation-and-geoinformation>, accessed 21 July 2022.

'Whales are vital to curb climate change - this is the reason why', World Economic Forum, 29 November 2019, <weforum.org/agenda/2019/11/whales-carbon-capture-climate-change/>.

F. Armstrong, A. Capon & R. McFarlane, 'Coronavirus is a wake-up call: our war with the environment is leading to pandemics', *The Conversation*, 31 March 2020, <theconversation.com/coronavirus-is-a-wake-up-call-our-war-with-the-environment-is-leading-to-pandemics-135023>.

G. Raygorodetsky, 'Indigenous peoples defend Earth's biodiversity – but they're in danger', *National Geographic*, 17 November 2018, <nationalgeographic.com/environment/article/can-indigenous-land-stewardship-protect-biodiversity->.

Chapter 10

K. Lagrave, 'Inside the exclusive club of travelers racing to visit every country in the world', *Condé Nast Traveler*, 26 April 2019, <cntraveler.com/story/inside-the-exclusive-club-of-travelers-racing-to-visit-every-country-in-the-world>.

References

C. Kahn, 'As "voluntourism" explodes in popularity, who's it helping most?', *NPR*, 31 July 2014, <npr.org/sections/goatsandsoda/2014/07/31/336600290/as-volunteerism-explodes-in-popularity-whos-it-helping-most>.

C. Platt, 'Alice Springs desert plane storage: Singapore Airlines A380s, 777s arrive in Australian desert', *Traveller*, 5 May 2020, <traveller.com.au/alice-springs-desert-plane-storage-singapore-airlines-a380s-777s-arrive-in-australian-desert-h1ntyv>.

'Workplace stretch break coaching tool', Wellnomics, <wellnomics.com/solutions/stretch-break-coaching-tool/>, accessed 21 July 2022.

A. Kosciolek, 'Do cruise ships have doctors?', *Cruise*, 4 August 2020, <cruise.blog/2020/08/do-cruise-ships-have-doctors>.

'FedEx packages may soon be delivered by self-flying planes', *The Economic Times*, 28 September 2020, <economictimes.indiatimes.com/news/international/business/fedex-packages-may-soon-be-delivered-by-self-flying-planes/articleshow/78361392.cms?from=mdr>.

A. Boyle, 'Year in space: Jeff Bezos and his billionaire rivals finally usher in the age of commercial spaceflight', *GeekWire*, 31 December 2021, <geekwire.com/2021/year-in-space-jeff-bezos-and-his-billionaire-rivals-finally-usher-in-the-age-of-commercial-spaceflight/>.

T. Machemer, 'Trailblazing pilot Wally Funk will go to space 60 years after passing her astronaut tests', *Smithsonian Magazine*, 2 July 2021, <smithsonianmag.com/smart-news/wally-funk-trailblazing-pilot-will-go-space-60-years-after-passing-astronaut-tests-180978108/>.

'Mojave Aerospace Ventures wins the competition that started it all', XPRIZE, <xprize.org/prizes/ansari/articles/mojave-aerospace-ventures-wins-the-competition>, accessed 21 July 2022.

'Mars analog habitat', *Wikipedia*, <en.wikipedia.org/wiki/Mars_analog_habitat>, accessed 21 July 2022.

Silent Running, motion picture, 1972.

D. King, 'Holidaying in war zones: intrepid or irresponsible?', *The Sydney Morning Herald*, 7 April 2018, <smh.com.au/world/asia/holidaying-in-war-zones-intrepid-or-irresponsible-20180402-p4z7fc.html>.

'Longest barefoot journey', Guinness World Records, <guinnessworldrecords.com/world-records/91413-longest-barefoot-journey>, accessed 21 July 2022.

'Fastest time to travel to all seven continents', Guinness World Records, <guinnessworldrecords.com/world-records/80877-fastest-time-to-travel-to-all-seven-continents>, accessed 21 July 2022.

V. Nereim, B. Bartenstein & M. Martin, 'Saudi prince's $500 billion "Neom" megaproject woos Wall Street', *Bloomberg*, 7 March 2022, <bloomberg.com/news/articles/2022-03-07/saudi-prince-s-500-billion-neom-megaproject-woos-wall-street>.

E. Gillespie, 'The Instagram influencers hired to rehabilitate Saudi Arabia's image', *The Guardian*, 12 October 2019, <theguardian.com/world/2019/oct/12/the-instagram-influencers-hired-to-rehabilitate-saudi-arabias-image>.

Chapter 11

Highlander, motion picture, 1986.

Eternals, motion picture, 2021.

'Ship of Theseus', *Wikipedia*, <en.wikipedia.org/wiki/Ship_of_Theseus>, accessed 22 July 2022.

Eternal Sunshine of the Spotless Mind, motion picture, 2004.

'Gorilla guide: where they live, diet, and conservation', Discover Wildlife, <discoverwildlife.com/animal-facts/mammals/facts-about-gorillas/>, accessed 22 July 2022.

'Life expectancy', *Wikipedia*, <en.wikipedia.org/wiki/Life_expectancy>, accessed 22 July 2022.

Psalm 90:10, *King James Bible*.

M. Limb, 'Covid-19: pandemic reduced life expectancy in most developed countries, study finds', *British Medical Journal*, 2021, vol. 375, no. 2750, <bmj.com/content/375/bmj.n2750>.

Vanilla Sky, motion picture, 2001.

S. Boseley, 'Great expectations: today's babies are likely to live to 100, doctors predict', *The Guardian*, 2 October 2009, <theguardian.com/society/2009/oct/02/babies-likely-to-live-to-100>.

World population ageing 2017: highlights, U.N. Department of Economic and Social Affairs, 2017, <un.org/en/development/desa/population/publications/pdf/ageing/WPA2017_Highlights.pdf>.

A. Taylor, 'Japan sets a new record number for people over 100 years old – and almost all are women', *The Washington Post*, 14 September 2018, <washingtonpost.com/world/2018/09/14/japan-sets-new-record-number-people-over-years-old-almost-all-are-women/>.

'Ageing societies and the looming pension crisis', The Organisation for Economic Co-operation and Development, <oecd.org/general/ageingsocietiesandthelooming pensioncrisis.htm>, accessed 22 July 2022.

The Dying Rooms, documentary, 1995.

'Equal Pay Day', Workplace Gender Equality Agency, <wgea.gov.au/the-gender-pay-gap/equal-pay-day>, accessed 22 July 2022.

'When women are empowered, all of society benefits – Migiro', *UN News*, 16 November 2007, <news.un.org/en/story/2007/11/239982-when-women-are-empowered-all-society-benefits-migiro>.

W.M. Carroll, 'The global burden of neurological disorders', *The Lancet*, 2019, vol. 8, no. 5, pp. 418–419, <thelancet.com/journals/laneur/article/PIIS1474-4422%2819%2930029-8/fulltext>.

D. Carrington, 'Climate "apocalypse" fears stopping people having children – study', *The Guardian*, 27 November 2020, <theguardian.com/environment/2020/nov/27/climate-apocalypse-fears-stopping-people-having-children-study>.

References

M. Buttigieg, 'Environmental impacts of funerals: burial vs cremation', blog, Bare., 11 March 2021, <bare.com.au/blog/environmental-impacts-of-funerals-death-burial-vs-cremation>.

J.R. Lee, 'My mushroom burial suit', video, TEDGlobal 2011, <ted.com/talks/jae_rhim_lee_my_mushroom_burial_suit>, accessed 22 July 2022.

'A quirky yet practical blog that explores the inevitable', The Bottom Drawer Book, <thebottomdrawerbook.com.au/blog/>, accessed 22 July 2022.

Chapter 12

'We shall fight on the beaches', International Churchill Society, <winstonchurchill.org/resources/speeches/1940-the-finest-hour/we-shall-fight-on-the-beaches/>, accessed 10 August 2022.

'Operation Biting, Bruneval, north France 27th/28th Feb 1942', Combined Operations, <combinedops.com/Bruneval.htm>, accessed 22 July 2022.

C. Linder, 'Skip the new Terminator, and watch this unnerving robo-apocalypse instead', Popular Mechanics, 28 October 2019, <popularmechanics.com/technology/robots/a29610393/robot-soldier-boston-dynamics/>.

B. Marr, 'The 4 Ds of robotization: dull, dirty, dangerous and dear', Forbes, 16 October 2017, <forbes.com/sites/bernardmarr/2017/10/16/the-4-ds-of-robotization-dull-dirty-dangerous-and-dear/?sh=71e57f233e0d>.

'Teletank', Wikipedia, <en.wikipedia.org/wiki/Teletank>, accessed 22 July 2022.

'Artificial hummingbird developed', Irish Independent, 18 February 2011, <independent.ie/world-news/and-finally/artificial-hummingbird-developed-26706343.html>.

Eye in the Sky, motion picture, 2015.

V. Koutroulis, 'Martens Clause', Oxford Bibliographies, reviewed 16 August 2017, <oxfordbibliographies.com/view/document/obo-9780199796953/obo-9780199796953-0101.xml>.

A. Wald, 'A method of estimating plane vulnerability based on damage of survivors', Center for Naval Analyses Alexandria Va. Operations Evaluation Group, 1 July 1980, <apps.dtic.mil/docs/citations/ADA091073>.

The Truman Show, motion picture, 1998.

The Matrix, motion picture, 1999.

Never Let Me Go, motion picture, 2010.

T. Poole, 'The myth and reality of the super soldier', BBC News, 8 February 2021, <bbc.com/news/world-55905354>.

'New McAfee report estimates global cybercrime losses to exceed $1 trillion', Business Wire, <businesswire.com/news/home/20201206005011/en/New-McAfee-Report-Estimates-Global-Cybercrime-Losses-to-Exceed-1-Trillion>, accessed 22 July 2022.

Z. Makelos Smith & E. Lostri, The hidden costs of cybercrime, McAfee, December 2020, <mcafee.com/enterprise/en-us/assets/reports/rp-hidden-costs-of-cybercrime.pdf>.

N. Cveticanin, 'What's on the other side of your inbox – 20 SPAM statistics for 2022', DataProt, updated 20 July 2022, <dataprot.net/statistics/spam-statistics/>.

L. Irwin, 'The cyber security risks of working from home', IT Governance, 19 August 2021, <itgovernance.co.uk/blog/the-cyber-security-risks-of-working-from-home>.

T. Brewster, 'Fraudsters cloned company director's voice in $35 million bank heist, police find', *Forbes*, 14 October 2021, <forbes.com/sites/thomasbrewster/2021/10/14/huge-bank-fraud-uses-deep-fake-voice-tech-to-steal-millions/?sh=d44e49775591>.

J. McCurry, 'South Korea silences loudspeakers that blast cross-border propaganda', *The Guardian*, 23 April 2018, <theguardian.com/world/2018/apr/23/south-korea-propaganda-machine-silence-summit>.

C. Gayomali, 'Trap streets: the crafty trick mapmakers use to fight plagiarism', *The Week*, 9 January 2015, <theweek.com/articles/466184/trap-streets-crafty-trick-mapmakers-use-fight-plagiarism>.

'Bird's-eye view could be key to navigating without GPS', U.S. Army DEVCOM Army Research Laboratory Public Affairs, 26 July 2021, <army.mil/article/248735/birds_eye_view_could_be_key_to_navigating_without_gps>.

G. Brumfiel, 'U.S. Navy brings back navigation by the stars for officers', *NPR*, 22 February 2016, <npr.org/2016/02/22/467210492/u-s-navy-brings-back-navigation-by-the-stars-for-officers>.

'Agreement governing the activities of states on the Moon and other celestial bodies', United Nations Office for Outer Space Affairs, <unoosa.org/oosa/en/ourwork/spacelaw/treaties/intromoon-agreement.html>, accessed 22 July 2022.

Gateway House, 'Who will control the Antarctic?', *Fair Observer*, 17 January 2013, <fairobserver.com/region/north_america/who-will-control-antarctic/>.

R. Sanders, 'Binary asteroid in Jupiter's orbit may be icy comet from solar system's infancy', *UC Berkeley News*, 1 February 2006, <berkeley.edu/news/media/releases/2006/02/01_patroclus.shtml>.

D. Gough & D. Cahn, 'Eritrean water targeted in war over arid land', *The Guardian*, 17 February 1999, <theguardian.com/world/1999/feb/17/8>.

'#Tech4Good, this hashtag we see everywhere', Use Design, <use.design/en/tech4good-this-hashtag-we-see-everywhere/>, accessed 22 July 2022.

'Live cyber threat map', Check Point, <threatmap.checkpoint.com>, accessed 22 July 2022.

Chapter 13

E. St. Vincent Millay, 'I, being born a woman and distressed', Poetry Foundation, <poetryfoundation.org/poems/148564/i-being-born-a-woman-and-distressed>, accessed 22 July 2022.

2001: A Space Odyssey, motion picture, 1968.

A. Taylor, '"Smartphone pinky" and other injuries caused by excessive phone use', *The Conversation*, 5 November 2020, <theconversation.com/smartphone-pinky-and-other-injuries-caused-by-excessive-phone-use-148861>.

'Brick Lane made Britain's first "Safe Text" street with padded lampposts to prevent mobile phone injuries', *Daily Mail*, updated 4 March 2008, <dailymail.co.uk/news/article-525785/Brick-Lane-Britains-Safe-Text-street-padded-lampposts-prevent-mobile-phone-injuries.html>.

D. Chaffey, 'Global social media statistics research summary 2022', Smart Insights, 1 June 2022, <smartinsights.com/social-media-marketing/social-media-strategy/new-global-social-media-research/>.

'August Kekulé', *Wikipedia*, <en.wikipedia.org/wiki/August_Kekulé>, accessed 22 July 2022.

Hashtag, motion picture, 2019.

W. Gibson, *Idoru*, Penguin, London, 1997.

Her, motion picture, 2013.

Blade Runner 2049, motion picture, 2017.

Ex Machina, motion picture, 2014.

A.I. Artificial Intelligence, motion picture, 2001.

Woebot Health, <woebothealth.com>, accessed 22 July 2022.

M. Greshko, 'Meet Sophia, the robot that looks almost human', *National Geographic*, 19 May 2018, <nationalgeographic.com/photography/article/sophia-robot-artificial-intelligence-science>.

R. Cairns, 'Meet Grace, the ultra-lifelike nurse robot', *CNN*, 19 August 2021 <edition.cnn.com/2021/08/19/asia/grace-hanson-robotics-android-nurse-hnk-spc-intl/index.html>.

'Australia media can be sued for social media comments, court rules', *BBC News*, 8 September 2021, <bbc.com/news/world-australia-58484205>.

A. Groth, 'You're the average of the five people you spend the most time with', *Business Insider*, 25 July 2012, <businessinsider.com/jim-rohn-youre-the-average-of-the-five-people-you-spend-the-most-time-with-2012-7>.

Chapter 14

E. Smart, *By Grand Central Station I Sat Down and Wept*, Editions Poetry, London, 1945.

'Weltschmerz', *Britannica*, 20 July 1998, <britannica.com/art/Weltschmerz>.

N. Rich, 'Losing Earth: the decade we almost stopped climate change', *The New York Times Magazine*, 1 August 2018, <nytimes.com/interactive/2018/08/01/magazine/climate-change-losing-earth.html>.

'Climate emergency declaration', *Wikipedia*, <en.wikipedia.org/wiki/Climate_emergency_declaration>, accessed 22 July 2022.

'Change management comic strips', Torben Rick, <torbenrick.eu/blog/change-management/change-management-comic-strips/>, accessed 22 July 2022.

D. Rice, '"Past a point of no return": reducing greenhouse gas emissions to zero still won't stop global warming, study says', *USA Today*, 12 November 2020, <usatoday.com/story/news/nation/2020/11/12/reducing-greenhouse-gas-emissions-stop-climate-change-study/3761882001/>.

'A whopping 91 percent of plastic isn't recycled', *National Geographic Resource Library*, <education.nationalgeographic.org/resource/whopping-91-percent-plastic-isnt-recycled>, accessed 22 July 2022.

'Fire at Coolaroo recycling plant under control, blankets Melbourne in smoke', *ABC News*, 28 February 2017, <abc.net.au/news/2017-02-28/fire-breaks-out-at-coolaroo-recycling-plant/8309168>.

N. Eckley Selin, 'carbon sequestration', *Britannica*, 22 July 2011, <britannica.com/technology/carbon-sequestration>.

'The nexus of biofuels, climate change, and human health: workshop summary', Roundtable on Environmental Health Sciences, Research, and Medicine; Board on Population Health and Public Health Practice; Institute of Medicine, Washington (D.C.), 2 April 2014.

'*Hindenburg* disaster', *Wikipedia*, <en.wikipedia.org/wiki/Hindenburg_disaster>, accessed 22 July 2022.

'Overview – Electric vehicles: tax benefits & purchase incentives in the European Union (2021)', ACEA, 24 November 2021, <acea.auto/fact/overview-electric-vehicles-tax-benefits-purchase-incentives-european-union-2021/>.

E. McPherson, 'The last country has stopped selling leaded petrol', *Nine News*, 31 August 2021, <9news.com.au/world/the-last-country-has-stopped-selling-leaded-petrol/7b171ca3-4527-44de-b03d-f40002d7385a>.

J. Gambrell, 'Massive cargo ship turns sideways, blocks Egypt's Suez Canal', *KARE*, 23 March 2021, <kare11.com/article/news/nation-world/cargo-ship-suez-canal-trapped/507-3ca6964c-3ac2-4b13-867c-38d87366ea5d>.

R. Beighton, 'World's first crewless, zero emissions cargo ship will set sail in Norway', *CNN*, updated 27 August 2021, <edition.cnn.com/2021/08/25/world/yara-birkeland-norway-crewless-container-ship-spc-intl/index.html>.

'All in a days work: Qantas zeroes in on emissions, announces new CMO and launches shortest international flight', *Travel Weekly*, 31 March 2022, <travelweekly.com.au/article/all-in-a-days-work-qantas-zeroes-in-on-emissions-announces-new-cmo-and-launches-shortest-international-flight/>.

A. Greig, 'Introducing Qantas' first landfill-free flight', *Travel Insider*, 8 May 2019, <qantas.com/travelinsider/en/trending/new-qantas-flight-waste-free-sustainable.html>.

J. Bouwer, V. Krishnan, S. Saxon & C. Tufft, 'Taking stock of the pandemic's impact on global aviation', McKinsey & Company, 31 March 2022, <mckinsey.com/industries/travel-logistics-and-infrastructure/our-insights/taking-stock-of-the-pandemics-impact-on-global-aviation>.

To dream the impossible dream

W. Shakespeare & G.R. Hibbard, *Hamlet*, Oxford University Press, Oxford, 1994.

'Edward Hodges Baily RA (1788–1867)', Royal Academy of Arts, <royalacademy.org.uk/art-artists/name/edward-hodges-baily-ra>, accessed 22 July 2022.

'The iconic think different Apple commercial narrated by Steve Jobs', *fs*, <fs.blog/steve-jobs-crazy-ones/>, accessed 22 July 2022.

Index

Index

Dr Cath never gives the same presentation
twice, with each conversation carefully curated
to excite, entertain, challenge, educate or
scare the wits out of your audience
(you can choose which style you'd prefer),
from boardroom briefings on emerging topics
of interest to project design and direction
via large conference events.

Get in touch to learn more at
www.DrCatherineBall.com.